PERGAMON INTERNATIONAL LIBRARY
of Science, Technology, Engineering and Social Studies
The 1000-volume original paperback library in aid of education, industrial training and the enjoyment of leisure
Publisher: Robert Maxwell, M.C.

Worked Examples in Turbomachinery
(Fluid Mechanics and Thermodynamics)

Other Titles of Interest in the Pergamon International Library

Benson	:	Advanced Engineering Thermodynamics
Bradshaw	:	Experimental Fluid Mechanics, 2nd Edition
Buckingham	:	The Laws and Applications of Thermodynamics
Dixon	:	Fluid Mechanics, Thermodynamics of Turbomachinery, 2nd Edition
Haywood	:	Analysis of Engineering Cycles, 2nd Edition
Morrill	:	An Introduction to Equilibrium Thermodynamics
Peerless	:	Basic Fluid Mechanics

Worked Examples in Turbomachinery

(Fluid Mechanics and Thermodynamics)

S.L.Dixon, Ph. D., B.Eng., M.I.Mech.E.,C.Eng.

Dept of Mechanical Engineering, University of Liverpool.

PERGAMON PRESS

OXFORD . NEW YORK . TORONTO
SYDNEY . PARIS . BRAUNSCHWEIG

Pergamon Press Offices:

U. K.	Pergamon Press Ltd., Headington Hill Hall, Oxford OX3 0BW, England
U. S. A.	Pergamon Press Inc., Maxwell House, Fairview Park, Elmsford, New York 10523, U.S.A.
CANADA	Pergamon of Canada, Ltd., 207 Queen's Quay West, Toronto 1, Canada
AUSTRALIA	Pergamon Press (Aust.) Pty. Ltd., 19a Boundary Street, Rushcutters Bay, N.S.W. 2011, Australia
FRANCE	Pergamon Press SARL, 24 rue des Ecoles, 75240 Paris, Cedex 05, France
WEST GERMANY	Pergamon Press GMbH, 3300 Braunschweig, Postfach 2923, Burgplatz 1, West Germany

First edition 1975

Library of Congress Cataloging in Publication Data

Dixon, Sydney Lawrence.
Worked examples in turbomachinery. (fluid mechanics and thermodynamics)

(Thermodynamics and fluid mechanics series) (Pergamon international library)
1. Turbomachines--Fluid dynamics--Problems, exercises, etc. 2. Thermodynamics--Problems, exercises, etc. I. Title.
TJ267.D52 1975 621.4'06 75-9757
ISBN 0-08-019700-0

Printed in Great Britain by Biddles Ltd., Guildford, Surrey
ISBN 0 08 019700 0

List of Contents

Editorial Foreword

The books in the Thermodynamics and Fluid Mechanics Series are a planned set of short texts each covering specific topics. They are now well established text-books for many Engineering Degree Courses and they also serve as introductory reading and updating material for Engineers in Industry.

The present volume is a new venture in the Series and the objective is to provide more self-help for the reader. This is particularly relevant at the present time with the increasing use of SI units and rapid technological change.

W.A.W.
February, 1975.

Preface

This book is a companion volume to Fluid Mechanics, Thermodynamics of Turbomachinery, Second Edition (FMTT2) and is primarily concerned with the detailed solutions of the unworked problems set in that volume. In most engineering courses there is usually only a limited amount of time available to deal with the technique of problem solving. In turbomachinery courses it is vital for the student not only to understand the theoretical development of the various expressions he meets but to learn to apply them to the numerical solution of problems. Some students, in fact, become better acquainted with the analytical aspects of their courses after a close study of the numerical applications of the relevant formulae. In FMTT2 about a dozen worked examples were included in the text but these were, necessarily, rather simple applications of the theoretical treatment. In the present text the problems solved are of a standard comparable to, and in some cases harder than, those set in examinations for an Honours Degree in Mechanical Engineering and should be an invaluable aid to students sitting this or similar examinations.

I have confined the number of theoretical derivations to a minimum and have referred fairly frequently to equations already derived in FMTT2. On a point of clarification and to avoid repetition, I have used roman numerals for referencing equations derived in the worked solutions but have used arabic numerals when citing equations given in FMTT2 (e.g. eqn. (2.1)).

In order to obtain numerical precision and to avoid the forward propagation of errors all the calculations have been done with an electronic calculator. The use of this increasingly popular aid revealed a number of discrepancies in the answers given in FMTT2 which were previously evaluated with a standard (ten inch) slide rule. Calculations of the type involving the small difference of two numbers, particularly when one or both of these numbers had been determined by the process of exponentiation, are a common source of error in the arithmetical solution of turbomachinery problems but, fortunately, readily avoided by recent advances in inexpensive solid-state circuitry! All the intermediate and final results are shown rounded up to four significant figures which is adequate in engineering calculations of this type.

A number of problems refer to the use of a Mollier chart for steam and to steam tables. The data referred to was obtained from,

(i) Enthalpy-Entropy Diagram for Steam (SI Units), prepared by D.C. Hickson and F.R. Taylor, (Blackwell),

(ii) Thermodynamic and Transport Properties of Fluids (SI Units), arranged by Y.R. Mayhew and G.F.C. Rogers, Second Edition, 1967, (Blackwell).

I am grateful to Mr. John Blackburn, B.Eng., for carefully checking the manuscript and numerical working to Mrs. Avril Bevan for her meticulous care with the typing of the "camera ready" copy and to Dr. William Woods for his general encouragement and useful suggestions during the preparation of the book. Last but by no means least I thank my wife, Rosaleen, for her patience with me and skill in keeping the younger generation under control while I worked.

S.L. DIXON
LIVERPOOL, 1975.

SYMBOLS

A	area
a	sonic velocity
b	passage width
C_f	tangential force coefficient
C_L, C_D	lift and drag coefficients
C_p	specific heat at constant pressure, pressure coefficient
C_v	specific heat at constant volume
c	absolute velocity
c_o	spouting velocity
D	drag force, diameter
D_{eq}	equivalent diffusion ratio
D_h	hydraulic mean diameter
E, e	energy, specific energy
f	acceleration
g	gravitational acceleration
H	head, blade height
h	specific enthalpy
i	incidence angle
K, k	constants
L	lift force
ℓ	blade chord length
M	Mach number
m	mass, molecular 'weight'
N	rotational speed
N_s	specific speed
N_{sp}	power specific speed
n	number of stages, polytropic index
p	pressure
Q	heat transfer, volume flow rate
q	dryness fraction
R	reaction, specific gas constant
Re	Reynolds number
R_H	reheat factor
R_o	Universal gas constant
r	radius
S	entropy
s	blade pitch, specific entropy
T	absolute temperature
t	time, thickness
U	blade speed, internal energy
u	specific internal energy
V, v	volume, specific volume
W	work transfer
ΔW	specific work transfer
w	relative velocity
X	axial force
x, y, z	coordinate directions
Y	tangential force
Y_p	profile loss coefficient
Z	number of blades, blade loading parameter
α	absolute flow angle
β	relative flow angle
Γ	circulation
γ	ratio of specific heats
δ	deviation angle
ε	fluid deflection angle, cooling effectiveness
ζ	enthalpy loss coefficient
η	efficiency
Θ	minimum opening
θ	blade camber angle, wake momentum thickness
λ	profile loss coefficient
μ	dynamic viscosity
ν	kinematic viscosity
ξ	blade stagger angle
ρ	density
σ	slip factor
τ	torque
ϕ	flow coefficient, velocity ratio
ψ	stage loading factor
Ω	angular speed of rotation
Ω_s	angular specific speed
ω	vorticity
$\bar{\omega}$	stagnation pressure loss coefficient

Subscripts

av	average
c	compressor, critical
D	diffuser
e	exit
h	hydraulic, hub
i	inlet, impeller

id	ideal
is	isentropic
m	mean, meridional, mechanical
N	nozzle
n	normal component
o	stagnation property, overall
p	polytropic, constant pressure
R	reversible process, rotor
r	radial
rel	relative
s	isentropic, stall condition
ss	stage isentropic
t	turbine, tip, transverse
v	velocity
x, y, z	coordinate components
θ	tangential

Superscripts

$\cdot$	time rate of change
$-$	average
$'$	blade angle (as distinct from flow angle)
*	nominal condition

Chapter I

Dimensional Analysis, Similitude

1.1 A fan operating at 1750 rev/min at a volume flow rate of 4.25 m^3/s develops a head of 153 mm measured on a water-filled U-tube manometer. It is required to build a larger, geometrically similar fan which will deliver the same head at the same efficiency as the existing fan, but at a speed of 1440 rev/min. Calculate the volume flow rate of the larger fan.

Solution. For geometrically similar fans the dependent variables gH (the net energy transfer) and the efficiency η are expressed in terms of two functional relationships of the independent variables,

$$gH = f_1(Q, N, D, \rho, \mu)$$
$$\eta = f_2(Q, N, D, \rho, \mu)$$

where Q is the volume flow rate, D is a characteristic diameter of a fan, N is the rotational speed, μ the dynamic viscosity of the fluid and ρ the fluid density. Using either the formal procedure of dimensional analysis or the less formal but more direct process of dimensional elimination (see Q.1.4) with ρ, N, D as common factors, the dimensionless groups are

$$\frac{gH}{N^2D^2} = f_1\left(\frac{Q}{ND^3}, \frac{\rho ND^2}{\mu}\right)$$
$$\eta = f_2\left(\frac{Q}{ND^3}, \frac{\rho ND^2}{\mu}\right)$$

The group $\rho ND^2/\mu$ defines a flow Reynolds number Re based upon blade speed and fan diameter. It is assumed for the purpose of this problem that the effects of changes in Re are small and can be ignored. Thus, the performance characteristics are reduced to

$$\frac{gH}{N^2D^2} = f_1\left(\frac{Q}{ND^3}\right) \qquad \text{(i)}$$
$$\eta = f_2\left(\frac{Q}{ND^3}\right) \qquad \text{(ii)}$$

For the two fans to have the same efficiency it follows from eqn. (ii) that the volume flow coefficient $Q/(ND^3)$ must be the same. Thus, the volume flow rate of the second fan is

$$Q_2 = Q_1(N_2/N_1)(D_2/D_1)^3 \qquad \text{(iii)}$$

Likewise, for the two fans to deliver the same head, then $gH/(ND)^2$ must be the same which follows from eqn. (i) and the fact that $Q/(ND^3)$ is fixed. Hence, with $H_1 = H_2$, then

$$N_1D_1 = N_2D_2 \qquad \text{(iv)}$$

Substituting eqn. (iv) into eqn. (iii)

$$Q_2 = Q_1(N_1/N_2)^2$$
$$= 4.25(1750/1440)^2$$
$$= \underline{6.277 \text{ m}^3/\text{s}}$$

It will be observed that the numerical value of the head developed is not actually used in solving this problem.

1.2. An axial flow fan 1.83 m diameter is designed to run at a speed of 1400 rev/min with an average axial air velocity of 12.2 m/s. A quarter scale model has been built to obtain a check on the design and the rotational speed of the model fan is 4200 rev/min. Determine the axial air velocity of the model so that dynamical similarity with the full-scale fan is preserved. The effects of Reynolds number change may be neglected.

A sufficiently large pressure vessel becomes available in which the complete model can be placed and tested under conditions of complete similarity. The viscosity of the air is independent of pressure and the temperature is maintained constant. At what pressure must the model be tested?

<u>Solution</u>. The volume flow rate $Q = c_x A \propto c_x D^2$ where c_x is the axial velocity. Thus, the volume flow coefficient $Q/(ND^3)$ can be replaced with $c_x/(ND)$. For dynamical similarity and ignoring changes in Reynolds number the axial velocity of the model is

$$c_{xm} = c_{xp}\frac{N_m D_m}{N_p D_p} = 12.2 \times 4200/(1400 \times 4)$$

$$= \underline{9.15 \text{ m/s}}$$

For complete similarity the Reynolds number of the model must equal that of the prototype. Thus,

$$Re_m = Re_p$$

$$\frac{\rho_m N_m D_m^2}{\mu_m} = \frac{\rho_p N_p D_p^2}{\mu_p} \qquad \text{(i)}$$

As the temperature remains constant $\mu_m = \mu_p$ and, from the gas law, $\rho \propto p$. Thus, eqn. (i) becomes

$$p_m N_m D_m^2 = p_p N_p D_p^2$$

$$\therefore p_m = p_p(N_p/N_m)(D_p/D_m)^2$$

$$= 1 \times (1400/4200)4^2$$

$$= \underline{5.33 \text{ atm.}}$$

1.3. A water turbine is to be designed to produce 27 MW when running at 93.7 rev/min under a head of 16.5 m. A model turbine with an output of 37.5 kW is to be tested under dynamically similar conditions with a head of 4.9 m. Calculate the model speed and scale ratio. Assuming a model efficiency of 88%, estimate the volume flow rate through the model.

It is estimated that the force on the thrust bearing of the full-size machine will be 7.0 GN. For what thrust must the model bearing be designed?

Solution. For geometrically similar hydraulic turbines the dependent variables are the power output P, the efficiency η and the volume flow rate Q. The independent variables are the speed of rotation N, the characteristic diameter D, the useful head H, the dynamic viscosity μ and the density ρ. The functional dependences are written as

$$P, \eta, Q = f(\rho, N, D, gH, \mu)$$

By the application of dimensional analysis the following non-dimensional groups are formed using ρ, N and D as the common dimensional factors to eliminate the dimensions of the other variables (see Q.1.4),

$$\frac{P}{\rho N^3 D^5},\ \eta,\ \frac{Q}{ND^3} = f\left(\frac{gH}{N^2 D^2}, \frac{\rho N D^2}{\mu}\right)$$

A remarkable feature of dimensional analysis is the ability to form a new non-dimensional group from any other two such groups provided that the total number of groups in the functional relationship(s) remains the same. Thus, by combining the power coefficient with the head coefficent to eliminate the diameter (which is not given in the problem) a new non-dimensional group, the power specific speed is formed, i.e.

$$\frac{P}{\rho N^3 D^5}\left(\frac{N^2 D^2}{gH}\right)^{2.5} = \frac{P N^5 D^5}{\rho N^3 D^5 (gH)^{2.5}} = \frac{P N^2}{\rho (gH)^{2.5}}$$

Taking the square root of the above expression, the power specific speed is,

$$N_{sp} = NP^{1/2} / \left(\rho^{1/2} (gH)^{5/4}\right)$$

Assuming that changes in Reynolds number have negligible effect upon the performance, the model (m) and prototype (p) will have the same N_{sp} when operating under dynamically similar conditions. Thus,

$$\frac{N_m P_m^{1/2}}{H_m^{5/4}} = \frac{N_p P_p^{1/2}}{H_p^{5/4}}$$

$$\therefore N_m = N_p (P_p/P_m)^{1/2} (H_m/H_p)^{5/4}$$

$$= 93.7 (27 \times 10^6 / 37.5 \times 10^3)^{1/2} (4.9/16.5)^{5/4}$$

$$= \underline{551.2 \text{ rev/min}}$$

The scale ratio is determined from the head coefficient which is the same for model and prototype, i.e.

$$\frac{gH_m}{(N_m D_m)^2} = \frac{gH_p}{(N_p D_p)^2}$$

$$\therefore D_m/D_p = (H_m/H_p)^{1/2} N_p/N_m = (4.9/16.5)^{1/2}\, 93.7/551.2$$

$$= 0.09264$$

i.e. the prototype is <u>10.8 times</u> bigger than the model.

The turbine efficiency is defined as

$$\eta = P/(\rho g Q H)$$

hence, the volume flow rate of the model turbine is

$$\begin{aligned} Q_m &= P_m/(\rho g H_m \eta_m) \\ &= 37.5 \times 10^3/(10^3 \times 9.81 \times 4.9 \times 0.88) \\ &= \underline{0.8865 \text{ m}^3/\text{s}} \end{aligned}$$

The thrust force X is a new variable which can be related dimensionally to the other known variables. As force is the product of pressure ρgH and area which is proportional to D^2, then a force coefficient can be defined as

$$\hat{X} = X/(\rho g H D^2)$$

For dynamical similarity this will be the same for both model and prototype, hence

$$\begin{aligned} X_m &= X_p(H_m/H_p)(D_m/D_p)^2 \\ &= 7 \times 10^9(4.9/16.5)(0.09264)^2 \\ &= \underline{17.84 \text{ MN}} \end{aligned}$$

1.4. Derive the non-dimensional groups that are normally used in the testing of gas turbines and compressors.

A compressor has been designed for normal atmospheric conditions (101.3 kPa and 15°C). In order to economise on the power required it is being tested with a throttle in the entry duct to reduce the entry pressure. The characteristic curve for its normal design speed of 4000 rev/min is being obtained on a day when the ambient temperature is 20°C. At what speed should the compressor be run? At the point on the characteristic curve at which the mass flow would normally be 58 kg/s the entry pressure is 55 kPa. Calculate the actual rate of mass flow during the test.

Describe the relationship between geometry and specific speed for pumps.

<u>Solution.</u> As the fluid density ρ can change very appreciably across compressors and gas turbines of large pressure ratio it is necessary to employ compressible fluid relations. For a compressor of a given configuration and size represented by

a diameter D, operating at rotational speed N and mass flow rate $\dot{m}$ and at specified inlet stagnation conditions (p_{o1}, T_{o1}), the dependent performance parameters are the outlet stagnation pressure p_{o2}, the efficiency η and the overall stagnation temperature rise ΔT_o. Other dependent parameters may also be used (c.f. eqn. (1.13)). The dependent variables can be expressed in the form of three unknown functional relationships as

$$p_{o2},\ \eta\ ,\ \Delta T_o \quad = \quad f(N, D, \dot{m}, p_{o1}, T_{o1}, \gamma, \mu)$$

where μ is the dynamic viscosity and $\gamma = C_p/C_v$.

The dependent variables can be made dimensionless without difficulty by writing

$$p_{o2}/p_{o1},\ \eta,\ \Delta T_o/T_{o1} \quad = \quad f(N, D, \dot{m}, \rho_{o1}, a_{o1}, \gamma, \mu)$$

where, for convenience, p_{o1} and T_{o1} have been replaced by $\rho_{o1} = p_{o1}/(RT_{o1})$ and $a_{o1} = (\gamma RT_{o1})^{1/2}$. The most convenient and least formal method of finding the remaining dimensionless groups is to take one variable of interest and reduce its dimensions to zero by repeated multiplication with several other variables. To do this, three of the most easily measured variables ρ_{o1}, N and D are selected. These variables have the respective dimensions ML^{-3}, T^{-1} and L. Considering in detail the reduction of $\dot{m}$ to a dimensionless group by repeated multiplication,

Variable	Dimensions	Eliminating
$\dot{m}$	MT^{-1}	
$\dot{m}/\rho_{o1}$	L^3T^{-1}	M
$\dot{m}/(\rho_{o1}D^3)$	T^{-1}	L
$\dot{m}/(\rho_{o1}ND^3)$	o	T

The same process is used to reduce the remaining dimensional variables $a_{o1} (\equiv LT^{-1})$ and $\mu (\equiv ML^{-1}T^{-1})$ resulting in

$$\frac{p_{o2}}{p_{o1}},\ \eta\ ,\ \frac{\Delta T_o}{T_{o1}} \quad = \quad f\left(\frac{\dot{m}}{\rho_{o1}ND^3},\ \frac{\rho_{o1}ND^2}{\mu},\ \frac{ND}{a_{o1}},\ \gamma\right)$$

The group $\rho_{o1}ND^2/\mu$ is a Reynolds number Re based upon blade speed ($\propto$ND) and compressor size. The group ND/a_{o1} is a blade Mach number M. The group

$\dot{m}/(\rho_{o1}ND^3)$ is not very convenient for compressor testing but may be easily transformed into normal form as follows:-

$$\frac{\dot{m}}{\rho_{o1}ND^3} = \frac{\dot{m}RT_{o1}}{p_{o1}D^2(ND)} = \frac{\dot{m}\sqrt{(RT_{o1})}}{p_{o1}D^2}\left(\frac{a_{o1}/\gamma^{1/2}}{(ND)}\right)$$

$$= \frac{\dot{m}\sqrt{(RT_{o1})}}{p_{o1}D^2}$$

as the group $\left(a_o/(ND\gamma^{1/2})\right) = \left(1/(M\gamma^{1/2})\right)$ is a combination of dimensionless variables which have already appeared as separate independent groups it can be simply deleted from the above group. Thus, the final non-dimensional form of the compressor functional relationships is

$$\frac{p_{o2}}{p_{o1}}, \eta, \frac{\Delta T_o}{T_{o1}} = f\left(\frac{\dot{m}(RT_{o1})^{1/2}}{p_{o1}D^2}, \quad Re, \quad M, \quad \gamma\right) \qquad \text{(i)}$$

For a given compressor of a given size and handling a specific gas it has become the custom in practice to drop γ, R and D from the above set of dimensionless groups. The resulting relationships are then

$$\frac{p_{o2}}{p_{o1}}, \eta, \frac{\Delta T_o}{T_{o1}} = f\left(\frac{\dot{m}T_{o1}^{1/2}}{p_{o1}}, \quad \frac{N}{T_{o1}^{1/2}}, \quad Re\right) \qquad \text{(ia)}$$

which are no longer dimensionless.

In the case of a turbine, the dependent variables are usually regarded as $\dot{m}$, η and ΔT_o so that the dimensional functional relationships are

$$\dot{m}, \eta, \Delta T_o = f(p_{o1}, p_{o2}, T_{o1}, N, D, \gamma, \mu)$$

By a process of reasoning similar to that used for a compressor the variables are reduced to a smaller number of non-dimensional groups, i.e.

$$\frac{\dot{m}(RT_{o1})^{1/2}}{p_{o1}D^2}, \eta, \frac{\Delta T_o}{T_{o1}} = f\left(\frac{ND}{a_{o1}}, \frac{p_{o1}}{p_{o2}}, \frac{\dot{m}}{D\mu}, \gamma\right)$$

Referring to Fig. 1.9, and ignoring any effects due to changes in Re, the design point of a compressor is uniquely represented by one value of $N/T_{01}^{1/2}$ and one value of $\dot{m}T_{01}^{1/2}/p_{01}$. At the intersection of these two curves there is but one value for each of the dependent variables, p_{02}/p_{01}, η and $\Delta T_0/T_{01}$. Under normal atmospheric conditions, $p_{01} = 101.3$ kPa and $T_{01} = 288$ K, the compressor design speed N is 4000 rev/min and the mass flow rate $\dot{m}$ is 58 kg/s. Design point conditions still obtain for the new entry conditions $p_{01}' = 55$ kPa, $T_{01}' = 293$ K by adjusting the speed and mass flow rate to maintain the design point values of $\dot{m}T_{01}^{1/2}/p_{01}$ and $N/T_{01}^{1/2}$. Thus,

$$N' = N(T_{01}'/T_{01})^{1/2} = 4000\,(293/288)^{1/2}$$

$$= \underline{4035 \text{ rev/min}}$$

$$\dot{m}' = (p_{01}'/p_{01})(T_{01}/T_{01}')^{1/2}\dot{m} = (55/101.3)(288/293)^{1/2} \times 58$$

$$= \underline{31.22 \text{ kg/s}}$$

For a pump specific speed is defined, eqn. (1.8), by

$$N_s = NQ^{1/2}/(gH)^{3/4}$$

where N is the speed of rotation, Q the volume flow rate and H the head rise. For a given speed N, high specific speed would be obtained in a pump of small head rise and large volume flow rate, e.g. an axial flow pump. Conversely, a low specific speed pump would be typified by a radial flow machine of relatively small flow rate and a high head rise.

Chapter 2

Thermodynamics

2.1. For the adiabatic expansion of a perfect gas through a turbine, show that the overall efficiency η_t and small stage efficiency η_p are related by

$$\eta_t = (1 - \varepsilon^{\eta_p})/(1 - \varepsilon)$$

where $\varepsilon = r^{(1-\gamma)/\gamma}$, and r is the expansion pressure ratio, γ is the ratio of specific heats.

An axial flow turbine has a small stage efficiency of 86%, an overall pressure ratio of 4.5 to 1 and a mean value of γ equal to 1.333. Calculate the overall turbine efficiency.

Solution. The overall efficiency of a turbine is assumed to mean the total to total efficiency defined, eqn. (2.21), by

$$\eta_t = (h_{o1} - h_{o2})/(h_{o1} - h_{o2s})$$

For a perfect gas, $h = C_p T$, so that

$$\eta_t = (T_{o1} - T_{o2})/(T_{o1} - T_{o2s})$$

$$= (1 - T_{o2}/T_{o1})/(1 - T_{o2s}/T_{o1}) \quad \ldots \text{(i)}$$

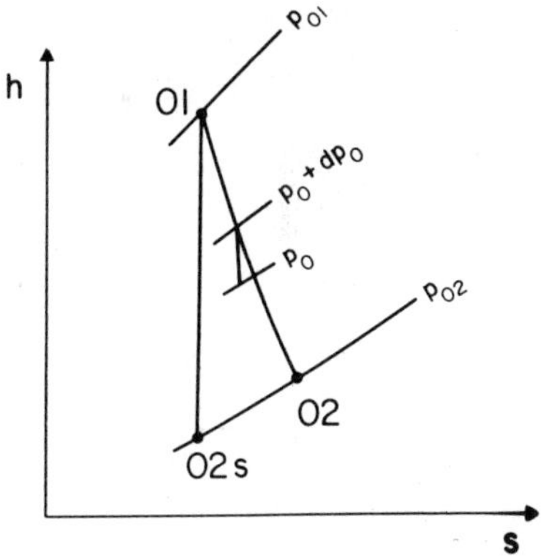

(ii)

The overall total pressure ratio is

$$r = p_{o1}/p_{o2} = (T_{o1}/T_{o2s})^{\gamma/(\gamma-1)}$$

$$\therefore \varepsilon = T_{o2s}/T_{o1} = r^{(1-\gamma)/\gamma}$$

Consider a small part of the expansion process as shown in the sketch. This expansion is best imagined as a small stage with an enthalpy drop dh_o and corresponding pressure drop dp_o. The small stage efficiency is defined as

$$\eta_p = dh_o/dh_{os}$$

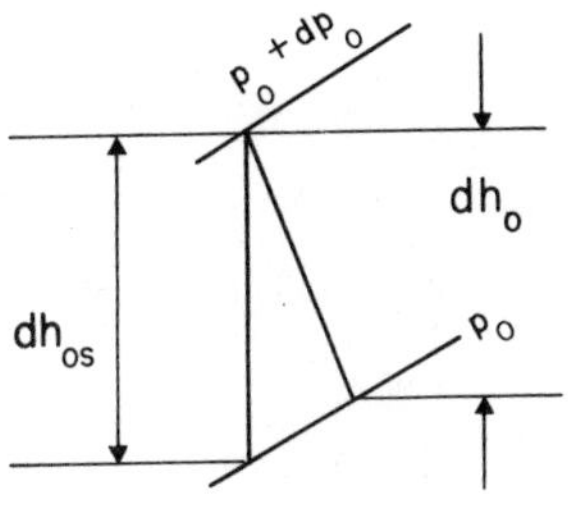

(iii)

Now an elementary change in specific entropy can be related, using the laws of thermodynamics, to the elementary changes in other properties. From eqn. (2.18)

$$Tds = dh - dp/\rho$$

which is applicable to both reversible and irreversible processes on a pure substance. For a constant entropy process it follows that

$$dh_{os} = dp_o/\rho_o$$

Substituting this result and the perfect gas relations, $p_o/\rho_o = RT_o$ and $dh_o = C_p\, dT_o$ into eqn. (iii)

$$\eta_p = \rho_o C_p\, dT_o/dp_o = (p_o/RT_o)\, \gamma R/(\gamma-1)\, dT_o/dp_o$$

where $C_p = \gamma R/(\gamma-1)$. After rearranging the above equation

$$dT_o/T_o = [\eta_p(\gamma-1)/\gamma]\, dp_o/p_o$$

Integrating and putting in the limits for the overall process,

$$T_{o2}/T_{o1} = (p_{o2}/p_{o1})^{\eta_p(\gamma-1)/\gamma}$$

$$= r^{\eta_p(1-\gamma)/\gamma} = \varepsilon^{\eta_p} \qquad \text{(iv)}$$

Substituting eqns. (ii) and (iv) into eqn. (i), the required relation is obtained,

$$\eta_t = (1 - \varepsilon^{\eta_p})/(1 - \varepsilon)$$

With $r = 4.5$, $\gamma = 1.333$ and $\eta_p = 0.86$,

$$\varepsilon = r^{(1-\gamma)/\gamma} = 1/4.5^{0.2498} = 1/1.456 = 0.6868.$$

$$\varepsilon^{\eta_p} = 0.6868^{0.86} = 0.7239$$

$$\therefore \eta_t = (1 - 0.7239)/(1 - 0.6868)$$

$$= \underline{88.16 \text{ per cent}}$$

2.2. Air is expanded in a multi-stage axial flow turbine, the pressure drop across each stage being very small. Assuming that air behaves as a perfect gas with ratio of specific heats γ, derive pressure-temperature relationships for the following processes:

(i) reversible adiabatic expansion;

(ii) irreversible adiabatic expansion, with small stage efficiency η_p;

(iii) reversible expansion in which the heat loss in each stage is a constant fraction k of the enthalpy drop in that stage;

(iv) reversible expansion in which the heat loss is proportional to the absolute temperature T.

Sketch the first three processes on a T, s diagram.

If the entry temperature is 1100 K, and the pressure ratio across the turbine is 6 to 1, calculate the exhaust temperatures in each of these three cases. Assume that γ is 1.333, that $\eta_p = 0.85$, and that k = 0.1.

Solution. (i) For a reversible adiabatic expansion the entropy does not change. From eqn. (2.18), with ds = 0,

$$Tds = dh - (1/\rho)dp = 0$$

$$\therefore dh = C_p\, dT = (1/\rho)dp = RT\,dp/p$$

$$\therefore dT/T = (R/C_p)dp/p = [(\gamma-1)/\gamma]\,dp/p$$

Integrating this result between limits, denoted by an initial state 1 and a final state 2, yields

$$\ln(T_1/T_2) = [(\gamma-1)/\gamma]\ \ln(p_1/p_2)$$

$$\therefore T_1/T_2 = (p_1/p_2)^{(\gamma-1)/\gamma} \qquad \text{(i)}$$

(ii) It has already been shown in the solution of Q.2.1 for an irreversible adiabatic expansion with small stage efficiency η_p, that

$$T_1/T_2 = (p_1/p_2)^{\eta_p(\gamma-1)/\gamma} \qquad \text{(ii)}$$

It is a consequence of the Second Law of Thermodynamics that the entropy of a substance (i.e. a system) undergoing an irreversible adiabatic process must increase. The magnitude of the entropy increase can be formulated from eqn. (2.18) as follows:-

$$Tds = dh - dp/\rho = C_p dT - RT\,dp/p$$

$$\therefore ds = C_p dT/T - R\,dp/p$$

Integrating and inserting limits, the entropy increase is

$$s_2 - s_1 = C_p \ln(T_2/T_1) - R \ln(p_2/p_1)$$
$$= R\{\ln(p_1/p_2) - [\gamma/(\gamma-1)] \ln(T_1/T_2)\}$$

Substituting for T_1/T_2 from eqn. (ii), and simplifying

$$s_2 - s_1 = R(1 - \eta_p) \ln(p_1/p_2) \qquad \text{(iia)}$$

(iii) From the Second Law of Thermodynamics, when an element of heat dQ_R is transferred reversibly from the surroundings to a unit mass of a substance at an absolute temperature T, the specific entropy increases by an amount

$$ds = dQ_R/T$$

Thus, a reversible heat transfer from the substance to the surroundings ($dQ_R < 0$) will cause the specific entropy to decrease. In the reversible expansion through the turbine with reversible heat loss the signs of the three elements ds, dh and dp in the expression $dQ_R = Tds = dh - dp/\rho$ are all negative. Writing $dQ_R = kdh$, eqn. (2.18) gives

$$Tds = kdh = dh - dp/\rho$$
$$\therefore (1-k)C_p dT = dp/\rho = RT\,dp/p$$
$$\therefore dT/T = (dp/p)(\gamma-1)/[\gamma(1-k)]$$

Integrating and inserting limits as before,

$$\therefore T_1/T_2 = (p_1/p_2)^{(\gamma-1)/[\gamma(1-k)]} \qquad \text{(iii)}$$

Additionally, it is easy to determine the magnitude of the corresponding specific entropy change, as follows:-

$$ds = kC_p dT/T$$
$$\therefore s_2 - s_1 = -kC_p \ln(T_1/T_2) = -kC_p(\gamma-1)/[\gamma(1-k)]\ln(p_1/p_2)$$
$$= -\ln(p_1/p_2)\,kR/(1-k) \qquad \text{(iiia)}$$

(iv) The heat loss in each elementary stage is reversible and proportional to T. This condition is satisfied by

$$dQ_R = Tds = dh - dp/\rho$$

$$\therefore\ ds = C_p\, dT/T - R\, dp/p$$

$$\therefore\ (s_2 - s_1)/R = \ln(p_1/p_2) - [\gamma/(\gamma-1)]\ln(T_1/T_2)$$

After rearranging and exponentiating,

$$p_1/p_2 = (T_1/T_2)^{\gamma/(\gamma-1)} . \exp[(s_2 - s_1)/R] \qquad \text{(iv)}$$

The sketch shows the way the entropy changes as the air is expanded from the initial state 1 to the final state 2 corresponding to the first three processes. The final temperatures (and specific entropy changes) are easily determined from the preceding equations.

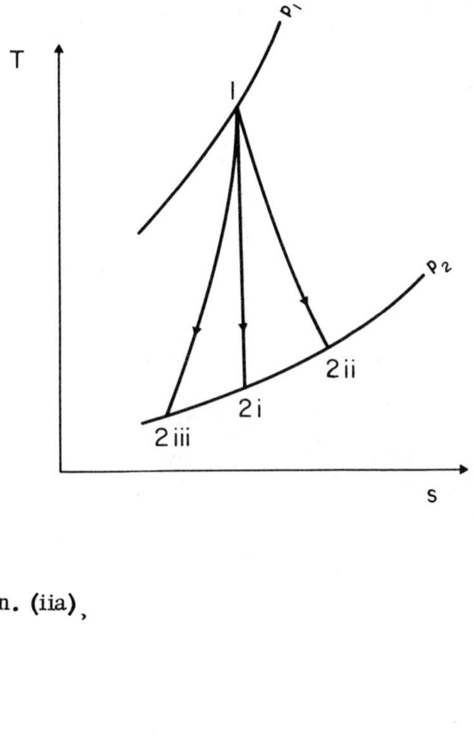

With $T_1 = 1100$ K, $p_1/p_2 = 6$, $\eta_p = 0.85$, $\gamma = 1.333$ and $k = 0.1$,

From eqn. (i),

$$T_{2i} = 1100/6^{0.2498} = \underline{703.1\text{ K}}$$

From eqn. (ii)

$$T_{2ii} = 1100/6^{0.2123} = \underline{751.9\text{ K}}$$

and the corresponding entropy increase is, from eqn. (iia),

$$(s_{2ii} - s_1)/R = 0.15 \ln 6 = 0.2688$$

From eqn. (iii)

$$T_{2iii} = 1100/6^{0.2776} = \underline{669.0\text{ K}}$$

and the corresponding entropy change is from eqn. (iiia)

$$(s_{2iii} - s_1)/R = -(0.1/0.9) \ln 6 = -0.2$$

2.3. A multi-stage high-pressure steam turbine is supplied with steam at a stagnation pressure of 7 MPa abs. and a stagnation temperature of 500°C. The corresponding specific enthalpy is 3410 kJ/kg. The steam exhausts from the turbine at a stagnation pressure of 0.7 MPa abs., the steam having been in a superheated condition throughout the expansion. It can be assumed that the steam behaves like a perfect gas over the range of the expansion and that $\gamma = 1.3$. Given that the turbine flow process has a small-stage efficiency of 0.82, determine

(i) the temperature and specific volume at the end of the expansion;

(ii) the reheat factor.

The specific volume of superheated steam is represented by $pv = 0.231(h - 1943)$, where p is in kPa, v is in m^3/kg and h is in kJ/kg.

Solution. (i) In the notation of Q.2.1 the actual temperature ratio across the turbine, from eqn. (2.37), is

$$T_{o1}/T_{o2} = (p_{o1}/p_{o2})^{\eta_p(\gamma-1)/\gamma}$$

where $\eta_p(\gamma-1)/\gamma = 0.82 \times 0.3/1.3 = 0.1892$ and $p_{o1}/p_{o2} = 10$

$$\therefore T_{o1}/T_{o2} = 10^{0.1892} = 1.5461$$

The inlet stagnation temperature $T_{o1} = 500 + 273 = 773$ K, hence the outlet stagnation temperature is

$$T_{o2} = 773/1.5461 = \underline{500\ \text{K}}$$

The specific volume v_{o2} corresponding to stagnation conditions at outlet is determined with the superheated steam relation $pv = 0.231(h - 1943)$ and the perfect gas law $pv = RT$. Combining these two equations,

$$T_{o2}/T_{o1} = (h_{o2} - 1943)/(h_{o1} - 1943)$$

$$\therefore h_{o2} - 1943 = (h_{o1} - 1943)T_{o2}/T_{o1} = (3410 - 1943)500/773$$

$$= 948.9$$

$$\therefore h_{o2} = 2891.9\ \text{kJ/kg}$$

$$\therefore v_{o2} = 0.231 \times 948.9/700$$

$$= \underline{0.3131\ m^3/kg}$$

(ii) The reheat factor is defined, eqn. (2.39), as

$$R_H = \eta_t/\eta_p$$

where the overall or total to total efficiency is

$$\eta_t = (T_{o1} - T_{o2})/(T_{o1} - T_{o2s})$$

$$= (1 - T_{o2}/T_{o1})/(1 - T_{o2s}/T_{o1}) = (1 - T_{o2}/T_{o1})/\left[1 - (p_{o2}/p_{o1})^{(\gamma-1)/\gamma}\right]$$

$$= (1 - 1/1.5461)/(1 - 1/10^{0.2308})$$

$$= 0.5461 \times 1.7013/(1.5461 \times 0.7013)$$

$$= 0.8568$$

$$\therefore R_H = 0.8568/0.82 = \underline{1.045}$$

2.4. A 20 MW back-pressure turbine receives steam at 4 MPa and 300°C, exhausting from the last stage at 0.35 MPa. The stage efficiency is 0.85, the reheat factor 1.04 and the external losses 2% of the isentropic enthalpy drop. Determine the rate of steam flow.

At the exit from the first stage nozzles the steam velocity is 244 m/s, specific volume 68.6 dm^3/kg, mean diameter 762 mm and steam exit angle 76 deg measured from the axial direction. Determine the nozzle exit height of this stage.

Solution. From the definition of reheat factor, eqn. (2.39), the turbine total to total efficiency can be immediately determined:-

$$\eta_t = \eta_p R_H = 0.85 \times 1.04 = 0.884.$$

Using the notation given in Q.2.1 the isentropic stagnation enthalpy drop, $h_{o1} - h_{o2s}$, can be determined using steam tables or, less accurately but more quickly, using a Mollier diagram for steam. From steam tables at p_{o1} = 4 MPa (40 bar) and $T_{o1} = 300^\circ C$ the initial steam condition is superheated (about 50°C of superheat) with h_{o1} = 2963 kJ/kg and s_{o1} = 6.364 kJ/kg °C). Inspection of the tables shows that at p_{o2} = 0.35 MPa (3.5 bar) the vapour saturation value of specific entropy $s_{go2} > s_{o1}$. This means the isentropic state point o2s is in the liquid-vapour phase. The dryness fraction q can be evaluated for point o2s,

$$q = (s_{o2} - s_{fo2})/(s_{go2} - s_{fo2})$$

$$= (6.364 - 1.727)/5.214 = 0.8893$$

Hence, the specific stagnation enthalpy at point o2s is

$$h_{o2s} = h_{fo2} + q(h_{go2} - h_{fo2})$$
$$= 584 + 0.8893 \times 2148 = 2494 \text{ kJ/kg}$$

Thus, the isentropic stagnation enthalpy drop is

$$h_{o1} - h_{o2s} = 2963 - 2494 = 469 \text{ kJ/kg}$$

As the total to total efficiency is known the actual stagnation enthalpy drop can be found, i.e.

$$h_{o1} - h_{o2} = \eta_t(h_{o1} - h_{o2s}) = 0.884 \times 469 = 414.6 \text{ kJ/kg}$$

and this is the specific work done by the steam, ΔW. The actual specific work delivered at the output shaft is less than this because of the mechanical losses. The shaft power delivered is

$$\dot{W}_t = \eta_m \dot{m} \Delta W$$

where η_m is the mechanical efficiency. Thus, the rate of mass flow

$$\dot{m} = \dot{W}_t/(\eta_m \Delta W)$$
$$= 20 \times 10^6/(0.98 \times 414.6 \times 10^3)$$
$$= \underline{49.22 \text{ kg/s}}$$

From the equation of continuity and assuming uniform flow at all radii,

$$\dot{m} = \rho A c_x = \rho \pi d_m h c \cos \alpha$$

Hence, the blade height is

$$h = \dot{m} v/(\pi d_m c \cos \alpha_1)$$
$$= 49.22 \times 0.0686/(\pi \times 0.762 \times 244 \times .2419) = 0.0239 \text{ mm}$$
$$= \underline{23.9 \text{ mm}}$$

2.5. Steam is supplied to the first stage of a five stage pressure-compounded steam turbine at a stagnation pressure of 1.5 MPa and a stagnation temperature of 350°C. The steam leaves the last stage at a stagnation pressure of 7.0 kPa with a corresponding dryness fraction of 0.95. By using a Mollier chart for steam and assuming that the stagnation state point locus is a straight line joining the initial and final states, determine

(i) the overall total to total efficiency and total to static efficiency assuming the steam enters the condenser with a velocity of 200 m/s,

(ii) the stagnation conditions between each stage assuming that each stage does the same amount of work,

(iii) the total to total efficiency of each stage,

(iv) the reheat factor based upon stagnation conditions.

Solution. In this problem it is more accurate to use steam tables to determine the overall conditions as the final pressure is specified. For the interstage calculations where the pressures are not specified, the solutions required are greatly facilitated by the use of a Mollier chart for steam with only a small loss in accuracy.

(i) From the tables at $p_{01} = 1.5$ MPa (15 bar) and $T_{01} = 350^{\circ}C$ the stagnation enthalpy $h_{01} = 3148$ kJ/kg and the stagnation entropy $s_{01} = 7.102$ kJ/(kg $^{\circ}$C). Referring to the sketch of the Mollier diagram, the exhaust total condition is the state point 06. The exhaust stagnation enthalpy is

$$\begin{aligned} h_{06} &= h_{f06} + q(h_{g06} - h_{f06}) \\ &= 163 + 0.95 \times 2409 = 2451.6 \text{ kJ/kg} \end{aligned}$$

where h_{f06} and h_{g06} are the liquid and vapour saturation enthalpies at $p_{06} = 7$ kPa. Hence, the actual specific work done across the whole turbine is,

$$\begin{aligned} \Delta W &= h_{01} - h_{06} = 3148 - 2451.6 \\ &= 696.4 \text{ kJ/kg} \end{aligned}$$

The corresponding isentropic stagnation enthalpy h_{06ss} at the exhaust pressure p_{06} is obtained by determining the dryness fraction q_s.

$$\begin{aligned} q_s &= (s_{01} - s_{f06})/(s_{g06} - s_{f06}) = (7.102 - 0.559)/7.715 \\ &= 0.8481 \end{aligned}$$

$$\therefore h_{06ss} = h_{f06} + q_s(h_{g06} - h_{f06}) = 163 + 0.8481 \times 2409$$

$$= 2206 \text{ kJ/kg}$$

Hence, the overall isentropic stagnation enthalpy drop which is also the overall ideal specific work is

$$\Delta W_{max} = h_{o1} - h_{o6ss} = 3148 - 2206 = 942 \text{ kJ/kg}$$

The overall total to total efficiency is, eqn. (2.21),

$$\eta_{tt} = \Delta W / \Delta W_{max} = 696.4/942$$

$$= \underline{73.94\%}$$

The overall total to static efficiency is, eqn. (2.22),

$$\eta_{ts} = \Delta W/(\Delta W_{max} + \tfrac{1}{2} c_{6s}^2) \simeq W/(W_{max} + \tfrac{1}{2} c_6^2)$$

where it is assumed that $c_{6s}^2/2 = c_6^2/2 = \frac{1}{2} \times 200^2 = 20$ kJ/kg

$$\therefore \eta_{ts} = 696.4/(942 + 20)$$

$$= \underline{72.4\%}$$

(ii) and (iii) The specific work is divided equally between the five stages so that the specific work done per stage is

$$\Delta h_o = \Delta W/5 = 139.3 \text{ kJ/kg}$$

A straight line is drawn on the Mollier chart between state points 01 and 06 and the intermediate stage state points determined. The following table shows the stagnation pressures, stagnation temperatures or dryness fractions, the actual and isentropic stagnation enthalpies for each stage. The individual isentropic stagnation enthalpy drops for each stage are shown (Δh_{os}) and the stage total to total efficiencies determined from $\eta_{tt} = \Delta h_o / \Delta h_{os}$.

It should be noted that the values of h_{os} shown in the table are obtained from the intersection of the isentrope at the beginning of a stage with the isobar at the end of that stage. The reason the stage efficiencies increase as the flow proceeds through the turbine is because of the reduced slope of the constant pressure lines at the lower pressures (and lower temperatures) causing Δh_{os} to reduce.

State point	p_o (kPa)	T_o (°C)	q	h_o (kJ/kg)	h_{os} (kJ/kg)	Δh_{os}	Stage effic. %
01	1500	350	-	3148	-		
						225.0	61.9
02	620	274	-	3008.7	2923		
						212.7	65.5
03	240	200	-	2869.4	2796		
						199.4	69.9
04	85	125	-	2730.1	2670		
						193.1	72.1
05	26	-	0.988	2590.8	2537		
						185.8	75.0
06	7	-	0.95	2451.5	2405		

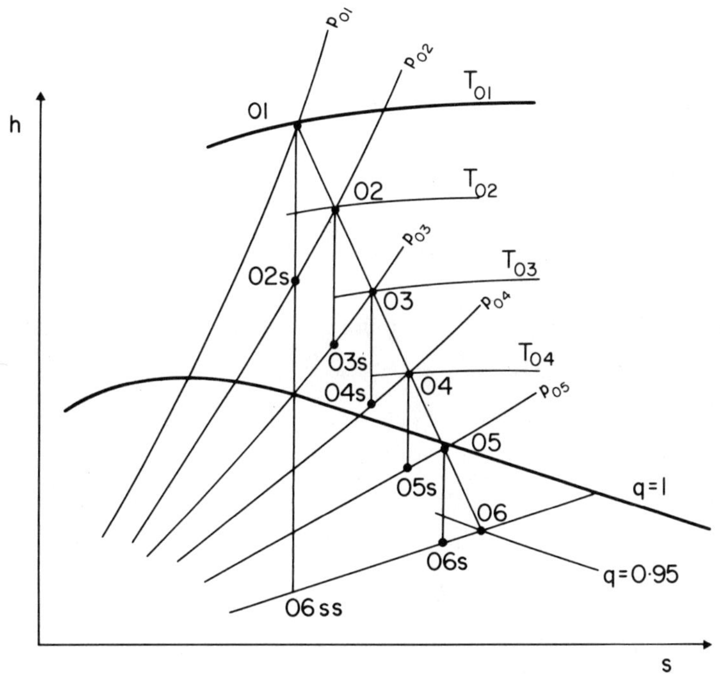

(iv) The reheat factor in a turbine is defined for a finite number of stages by

$$R_H = [(h_{o1} - h_{o2s}) + (h_{o2} - h_{o3s}) + \ldots\ldots] / (h_{o1} - h_{o6ss})$$

$$= \Sigma \Delta h_{os} / \Delta W_{max}$$

$$\therefore R_H = (225 + 212.7 + 199.4 + 193.1 + 185.8)/942$$

$$= \underline{1.078}$$

Chapter 3

Two-dimensional Cascades

3.1. Experimental compressor cascade results suggest that the stalling lift coefficient of a cascade blade may be expressed as

$$C_L\left(\frac{c_1}{c_2}\right)^3 = 2.2$$

where c_1 and c_2 are the entry and exit velocities. Find the stalling inlet angle for a compressor cascade of space-chord ratio unity if the outlet air angle is 30 deg.

Solution. The lift coefficient C_L of a cascade blade is defined, eqn. (3.16a), as the force L per unit blade length acting in the direction normal to the average velocity c_m divided by the product of the average dynamic pressure and blade chord ℓ. i.e.

$$C_L = L/(\tfrac{1}{2}\rho c_m^2 \ell)$$

where $c_m = c_x/\cos\alpha_m$ and $\tan\alpha_m = \frac{1}{2}(\tan\alpha_1 + \tan\alpha_2)$. For a compressor blade cascade C_L can be expressed, eqn. (3.18), in terms of the inlet flow angle α_1, the outlet flow angle α_2 the space/chord ratio s/ℓ and the drag coefficient C_D, as

$$C_L = 2(s/\ell)\cos\alpha_m(\tan\alpha_1 - \tan\alpha_2) - C_D\tan\alpha_m \qquad \text{(i)}$$

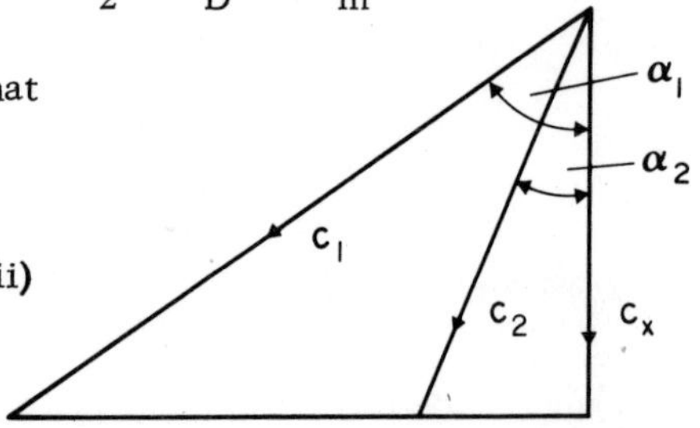

It is assumed in this problem that $C_D = 0$ and that c_x is constant. From the velocity triangles,

$$c_x = c_1\cos\alpha_1 = c_2\cos\alpha_2 \qquad \text{(ii)}$$

Thus, with eqns. (i) and (ii) and the given expression for stalling,

$$C_L(c_1/c_2)^3 = 2(s/\ell)\cos\alpha_m(\tan\alpha_1 - \tan\alpha_2)(\cos\alpha_2/\cos\alpha_1)^3 = 2.2$$

.... (iii)

With $s/\ell = 1.0$ and $\alpha_2 = 30$ deg, only α_1 remains unknown in eqn. (iii). Although it is possible to produce a polynomial equation in $\tan\alpha_1$ from eqn. (iii) it is far less trouble to solve the equation numerically. The procedure used is to select several values of α_1 and calculate the numerical values of α_m etc. until a value of α_1 satisfies the equation,

$$\cos\alpha_m(\tan\alpha_1 - \tan\alpha_2)/\cos^3\alpha_1 = 1.1/\cos^3\alpha_2 = 1.6936 \qquad \text{(iv)}$$

where $\tan\alpha_2 = 0.5774$, $\cos^3\alpha_2 = 0.6495$.

α_1°	45	47.5	49	50
$\tan\alpha_1$	1.0000	1.0913	1.1504	1.1918
$\tan\alpha_m$	0.7887	0.8343	0.8639	0.8846
α_m°	38.26	39.84	40.82	41.49
$\cos\alpha_m$	0.7852	0.7679	0.7567	0.7490
LHS eqn. (iv)	0.9387	1.280	1.5356	1.7320

By graphical interpolation the inlet flow stalling angle is

$$\alpha_1 = \underline{49.81 \text{ deg.}}$$

3.2. Show, for a turbine cascade, using the angle notation of Fig. 3.24, that the lift coefficient is

$$C_L = 2(s/\ell)(\tan\alpha_1 + \tan\alpha_2)\cos\alpha_m + C_D\tan\alpha_m$$

where $\tan\alpha_m = \frac{1}{2}(\tan\alpha_2 - \tan\alpha_1)$ and $C_D = \text{Drag}/(\frac{1}{2}\rho c_m^2\ell)$.

A cascade of turbine nozzle vanes has a blade inlet angle $\alpha_1 = 0$ deg, a blade outlet angle α_2' of 65.5 deg, a chord length ℓ of 45 mm and an axial chord b of 32 mm. The flow entering the blades is to have zero incidence and an estimate of the deviation angle based upon similar cascades is that δ will be about 1.5 deg at low outlet Mach number. If the blade load ratio ψ_T defined by eqn. (3.51) is to be 0.85, estimate a suitable space/chord ratio for the cascade.

Determine the drag and lift coefficients for the cascade given that the profile loss coefficient

$$\lambda = \Delta p_o/(\tfrac{1}{2}\rho c_2^2) = 0.035.$$

<u>Solution.</u> The figure shows part of a turbine blade cascade, the velocity triangle

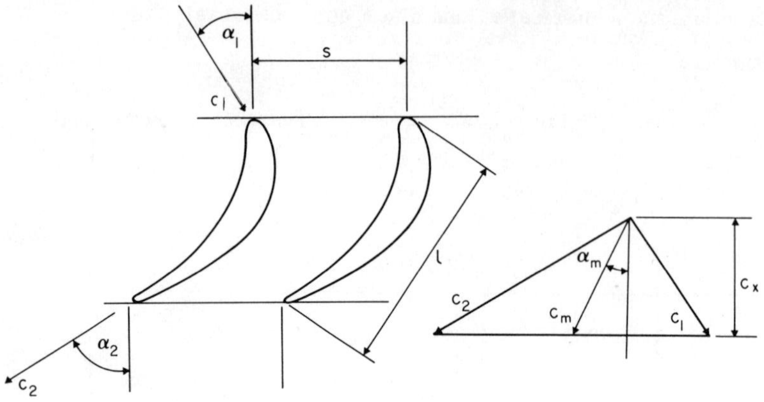

assuming c_x is constant and the force diagram. From the velocity triangle the mean flow direction α_m is defined by $\tan \alpha_m = \frac{1}{2}(\tan \alpha_2 - \tan \alpha_1)$ so that

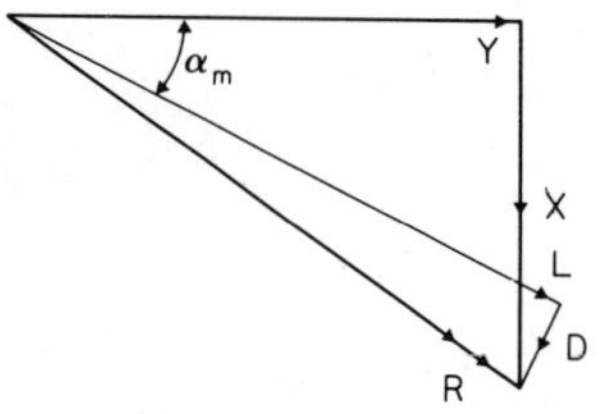

$c_m = c_x/\cos \alpha_m$. Referring to unit depth (span) of blade, the lift force L acting on the blade is perpendicular to c_m and the drag force D acting on the blade is parallel to c_m. The resultant force R has components X and Y in the axial and 'tangential' directions respectively. Resolving forces,

$$L = Y \cos \alpha_m + X \sin \alpha_m \qquad \text{(i)}$$

$$D = X \cos \alpha_m - Y \sin \alpha_m \qquad \text{(ii)}$$

With constant c_x the axial force acting on one blade is

$$X = (p_1 - p_2)s \qquad \text{(iii)}$$

The tangential force acting on one blade is, from the momentum equation,

$$Y = \rho s c_x(c_{y2} + c_{y1}) = \rho s c_x^2(\tan \alpha_2 + \tan \alpha_1) \qquad \text{(iv)}$$

where ρ is a mean density through the cascade. With the 'incompressible' flow approximation (for simplicity), $p_o = p + \frac{1}{2}\rho c^2$, then the total pressure loss across the cascade is,

$$\Delta p_o = p_{o1} - p_{o2} = p_1 - p_2 + \frac{1}{2}\rho(c_1^2 - c_2^2)$$

$$\therefore p_1 - p_2 = \Delta p_o - \frac{1}{2}\rho(c_1^2 - c_2^2) \qquad \text{(v)}$$

Substituting eqn. (v) into eqn. (iii),

$$X = s\,\Delta p_o + \frac{1}{2}\rho c_x^2(\sec^2\alpha_2 - \sec^2\alpha_1)s$$

$$= s\,\Delta p_o + \frac{1}{2}\rho c_x^2(\tan^2\alpha_2 - \tan^2\alpha_1)s$$

$$= s\,\Delta p_o + \rho s c_x^2 \tan\alpha_m(\tan\alpha_2 + \tan\alpha_1) \qquad \text{(vi)}$$

After using eqns. (iv) and (vi) in eqns. (i) and (ii) it follows that

$$D = s\,\Delta p_o \cos\alpha_m$$

$$L = \rho s c_x^2 \sec\alpha_m(\tan\alpha_2 + \tan\alpha_1) + s\,\Delta p_o \sin\alpha_m$$

With the definitions $C_L = L/(\frac{1}{2}\rho c_m^2 \ell)$ and $C_D = D/(\frac{1}{2}\rho c_m^2 \ell)$,

$$C_L = 2(s/\ell)\cos\alpha_m(\tan\alpha_2 + \tan\alpha_1) + C_D\tan\alpha_m \qquad \text{(vii)}$$

$$C_D = 2(s/\ell)\cos\alpha_m\,[\Delta p_o/(\rho c_m^2)] \qquad \text{(viii)}$$

The blade load ratio, eqn. (3.51), is

$$\psi_T = 2(s/b)\cos^2\alpha_2(\tan\alpha_2 + \tan\alpha_1)$$

At cascade exit the flow angle α_2 is less than the blade outlet angle α_2' by the amount of the deviation.

$$\alpha_2 = \alpha_2' - \delta = 65.5 - 1.5 = 64 \text{ deg}$$

At cascade inlet the blade angle α_1' is zero, the flow incidence is zero so that the flow angle $\alpha_1 = 0$. Thus, with $\alpha_1 = 0$, the space/chord ratio is

$$s/\ell = (b/\ell)(s/b) = (b/\ell)\psi_T/\sin 2\alpha_2$$

$$= (32/45)0.85/\sin(2 \times 64^\circ) = \underline{0.767}$$

From the velocity triangles, $c_x = c_2\cos\alpha_2 = c_m\cos\alpha_m$, then $c_m = c_2\cos\alpha_2/\cos\alpha_m$ and $\tan\alpha_m = \frac{1}{2}\tan\alpha_2$. Thus, $\alpha_m = 45.71^\circ$. Using this expression in eqn. (viii) the drag coefficient becomes

$$C_D = (s/\ell)\cos\alpha_m \left(\frac{\Delta p_0}{\frac{1}{2}\rho c_2^2}\right)\left(\frac{\cos\alpha_m}{\cos\alpha_2}\right)^2$$

$$= (s/\ell)\,\lambda\cos^3\alpha_m/\cos^2\alpha_2$$

$$= 0.767 \times 0.035 \times \cos^3 45.71^\circ/\cos^2 64^\circ$$

$$= \underline{0.0476}$$

From eqn. (vii) the lift coefficient can now be calculated

$$C_L = 2 \times 0.767 \times \cos 45.71^\circ \times \tan 64^\circ + 0.0476 \times \tan 45.71^\circ$$

$$= 2.196 + 0.049 = \underline{2.245}$$

N.B. In a turbine cascade with $\alpha_m > 0$, the drag slightly increases the lift which is the converse of what occurs in a compressor cascade.

3.3. A compressor cascade is to be designed for the following conditions:

Nominal fluid outlet angle	α_2^*	=	30 deg
Cascade camber angle	θ	=	30 deg
Pitch/chord ratio	s/ℓ	=	1.0

Using Howell's curves and his formula for nominal deviation, determine the nominal incidence, the actual deviation for an incidence of +2.7 deg and the approximate lift coefficient at this incidence.

Solution. The nominal deviation angle, eqn. (3.39) is

$$\delta^* = m\,\theta(s/\ell)^{1/2}$$

where, from eqn. (3.40a), the coefficient m is

$$m = 0.23(2a/\ell)^2 + \alpha_2^*/500$$

Assuming a circular arc camber line, $a/\ell = 0.5$, and

$$m = 0.23 + 30/500 = 0.29$$

$$\therefore \delta^* = 0.29 \times 30 \times 1 = 8.7 \text{ deg.}$$

Referring to the notation given in the sketch, the blade angles are,

$$\alpha_2' = \alpha_2^* - \delta^* = 30 - 8.7 = 21.3 \text{ deg}$$

$$\alpha_1' = \alpha_2' + \theta = 21.3 + 30 = 51.3 \text{ deg}$$

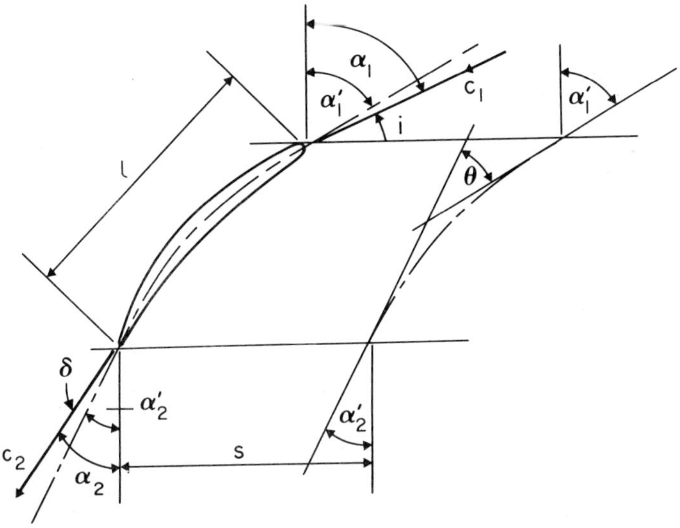

The nominal flow inlet angle can be obtained from the tangent difference approximation, eqn. (3.38), or less precisely from Fig. 3.16,

$$\tan \alpha_1^* = \tan \alpha_2^* + 1.55/[1 + 1.5(s/\ell)]$$

$$= \tan 30^{\circ} + 1.55/2.5 = 1.197$$

$$\therefore \alpha_1^* = 50.13 \text{ deg and } \varepsilon^* = \alpha_1^* - \alpha_2^* = 20.13 \text{ deg}$$

The nominal incidence is

$$i^* = \alpha_1^* - \alpha_1' = 50.13 - 51.3 = \underline{-1.17 \text{ deg}}$$

For $i = 2.7$ deg, $(i - i^*)/\varepsilon^* = (2.7+1.7)/20.13 = 0.190$. From Howell's curve of relative deflection $\varepsilon/\varepsilon^*$ against relative incidence $(i - i^*)/\varepsilon^*$, Fig. 3.17, the value of $\varepsilon/\varepsilon^* \doteqdot 1.15$. Hence, the actual deflection $\varepsilon = 1.15 \times 20.13 = 23.15$ deg. The actual inlet flow angle is $\alpha_1 = \alpha_1' + i = 51.3 + 2.7 = 54$ deg. The actual outlet flow angle is $\alpha_2 = \alpha_1 - \varepsilon = 54 - 23.15 = 30.85$ deg. Thus, the actual deviation angle for an incidence of 2.7 deg, is

$$= \alpha_2 - \alpha_2' = 30.85 - 21.3$$

$$= \underline{9.55 \text{ deg}}$$

The approximate lift coefficient, eqn. (3.17), is

$$C_L = 2(s/\ell)\cos\alpha_m(\tan\alpha_1 - \tan\alpha_2)$$

where it is assumed that C_D is negligible. The mean flow angle is

$$\tan\alpha_m = \frac{1}{2}(\tan\alpha_1 + \tan\alpha_2) = \frac{1}{2}(\tan 54^\circ + \tan 30.85^\circ) = 0.9868$$

$$\therefore \alpha_m = 44.62 \text{ deg}$$

$$\therefore C_L = 2 \times \cos 44.62^\circ(\tan 54^\circ - \tan 30.85^\circ)$$

$$= \underline{1.109}$$

3.4. A compressor cascade is built with blades of circular arc camber line, a space/chord ratio of 1.1 and blade angles of 48 and 21 deg at inlet and outlet. Test data taken from the cascade shows that at zero incidence (i = 0) the deviation $\delta = 8.2$ deg and the total pressure loss coefficient $\bar{\omega} = \Delta p_o/(\frac{1}{2}\rho c_1^2) = 0.015$. At positive incidence over a limited range $(0 \leqslant i \leqslant 6^\circ)$ the variation of both δ and $\bar{\omega}$ for this particular cascade can be represented with sufficient accuracy by linear approximations, viz.

$$\frac{d\delta}{di} = 0.06, \qquad \frac{d\bar{\omega}}{di} = 0.001$$

where i is in degrees.

For a flow incidence of 5.0 deg determine

(i) the flow angles at inlet and outlet;

(ii) the diffuser efficiency of the cascade;

(iii) the static pressure rise of air with a velocity 50 m/s normal to the plane of the cascade.

Assume the density of air is 1.2 kg/m^3.

Solution. (i) At zero incidence, i = 0, the deviation $\delta = \delta_o = 8.2$ deg and the total pressure loss coefficient $\bar{\omega} = \bar{\omega}_o = 0.015$. At i = 5 deg,

$$\delta = \delta_o + (d\delta/di)i = 8.2 + 0.06 \times 5 = 8.5 \text{ deg}$$

$$\bar{\omega} = \bar{\omega}_o + (d\bar{\omega}/di)i = 0.015 + 0.001 \times 5 = 0.02 .$$

The flow angles at 5 deg incidence are

$$\alpha_1 = \alpha_1' + i = 48 + 5 = \underline{53 \text{ deg}}$$

$$\alpha_2 = \alpha_2' + \delta = 21 + 8.5 = \underline{29.5 \text{ deg}}$$

(ii) The compressor cascade decelerates the flow between inlet and outlet and the efficiency of the process, assuming incompressible flow, can be expressed by the diffuser efficiency, eqn. (2.48),

$$\eta_D = (p_2 - p_1)/[\tfrac{1}{2}\rho(c_1^2 - c_2^2)]$$

As $p_2 - p_1 = p_{02} - p_{01} + \tfrac{1}{2}\rho(c_1^2 - c_2^2) = -\Delta p_o + \tfrac{1}{2}\rho(c_1^2 - c_2^2)$, then

$$\eta_D = 1 - \Delta p_o/[\tfrac{1}{2}\rho(c_1^2 - c_2^2)]$$

$$= 1 - \Delta p_o/[\tfrac{1}{2}\rho c_1^2(1 - \cos^2\alpha_1/\cos^2\alpha_2)]$$

$$= 1 - \bar{\omega}/(1 - \cos^2\alpha_1/\cos^2\alpha_2)$$

where $\bar{\omega} = \Delta p_o/(\tfrac{1}{2}\rho c_1^2)$ and $c_1 \cos\alpha_1 = c_2 \cos\alpha_2 = c_x$ are used.

Substituting values for α_1, α_2 and $\bar{\omega}$,

$$\eta_D = 1 - 0.02/(1 - \cos^2 53^\circ/\cos^2 29.5^\circ)$$

$$= \underline{0.962}$$

(iii) The static pressure rise is

$$p_2 - p_1 = \eta_D\,\rho(c_1^2 - c_2^2)/2 = \eta_D\,\rho c_x^2(\sec^2\alpha_1 - \sec^2\alpha_2)/2$$

$$= \eta_D\,\rho c_x^2(\tan^2\alpha_1 - \tan^2\alpha_2)/2$$

$$= 0.962 \times 1.2 \times 50^2(\tan^2 53^\circ - \tan^2 29.5^\circ)/2$$

$$= \underline{2.079 \text{ kPa}}$$

3.5.(a) A cascade of compressor blades is to be designed to give an outlet air angle α_2 of 30 deg for an inlet air angle α_1 of 50 deg measured from the normal to the plane of the cascade. The blades are to have a parabolic arc camber line with $a/\ell = 0.4$ (i.e. the fractional distance along the chord to the point of maximum camber). Determine the space/chord ratio and blade outlet angle if the cascade is to operate at zero incidence and nominal conditions. You may assume the linear approximation for nominal deflection of Howell's cascade correlation:

$$\varepsilon^* = (16 - 0.2\,\alpha_2^*)(3 - s/\ell)\ \text{deg}$$

as well as the formula for nominal deviation:

$$\delta^* = \left[0.23\left(\frac{2a}{\ell}\right)^2 + \frac{\alpha_2^*}{500}\right]\theta\sqrt{\frac{s}{\ell}}\ \text{deg}$$

(b) The space/chord ratio is now changed to 0.8, but the blade angles remain as they are in part (a) above. Determine the lift coefficient when the incidence of the flow is 2.0 deg. Assume that there is a linear relationship between $\varepsilon/\varepsilon^*$ and $(i - i^*)/\varepsilon^*$ over a limited region, viz. at $(i - i^*)/\varepsilon^* = 0.2$, $\varepsilon/\varepsilon^* = 1.15$ and at $i = i^*$, $\varepsilon/\varepsilon^* = 1$. In this region take $C_D = 0.02$.

Solution. (a) As the cascade is designed to operate at the 'nominal' condition, then the air angles given are also the nominal flow angles, i.e. $\alpha_1 = \alpha_1^* = 50$ deg and $\alpha_2 = \alpha_2^* = 30$ deg. Thus, the nominal deflection is

$$\varepsilon^* = \alpha_1^* - \alpha_2^* = (16 - 0.2\,\alpha_2^*)(3 - s/\ell) = 20\ \text{deg}$$

$$\therefore 20 = (16 - 0.2 \times 30)(3 - s/\ell) = 10(3 - s/\ell)$$

$$\therefore \underline{s/\ell = 1.0}$$

The nominal deviation is

$$^*\delta = \left[0.23(2a/\ell)^2 + \alpha_2^*/500\right]\theta\,(s/\ell)^{1/2}$$

$$= \left[0.23 \times 0.64 + 30/500\right]\theta = 0.2072\,\theta$$

As the incidence is zero the blade inlet angle $\alpha_1' = \alpha_1^* = 50$ deg. The nominal deflection is used again to solve for the blade camber, i.e.

$$\varepsilon^* = \alpha_1^* - \alpha_2^* = \alpha_1' - \alpha_2' - \delta^* = \theta - \delta^* = \theta(1 - 0.2072)$$

$$\therefore \theta = 20/(1 - 0.2072) = 25.2\ \text{deg}$$

Hence, the blade outlet angle is obtained from

$$\alpha_2' = \alpha_1' - \theta = 50 - 25.2$$

$$= \underline{24.8\ \text{deg}}$$

(b) The change to a smaller space/chord ratio will affect the nominal deviation and nominal flow outlet angles. The new nominal deviation is

$$\delta^* = (0.23 \times 0.64 + \alpha_2^*/500)\, 25.2\,(0.8)^{1/2}$$
$$= 3.318 + 0.04508\,\alpha_2^*$$

and the new nominal outlet angle is obtained from

$$\alpha_2^* = \alpha_2' + \delta^* = 24.8 + 3.318 + 0.04508\,\alpha_2^*$$
$$\therefore \alpha_2^* = 28.12/0.9549 = 29.45 \text{ deg}$$

The new nominal deflection is

$$\varepsilon^* = (16 - 0.2 \times 29.45)(3 - 0.8) = 22.24 \text{ deg}$$

Thus, the corresponding nominal inlet angle is

$$\alpha_1^* = \alpha_2^* + \varepsilon^* = 29.45 + 22.24 = 51.69 \text{ deg}$$

and the nominal incidence is obtained,

$$i^* = \alpha_1^* - \alpha_1' = 51.69 - 50 = 1.69 \text{ deg}$$

The linear relationship between deflection and incidence is in the form,

$$\varepsilon/\varepsilon^* - 1 = k(i - i^*)/\varepsilon^*$$

which satisfies the initial condition, i.e. $\varepsilon = \varepsilon^*$ when $i = i^*$. With $\varepsilon/\varepsilon^* = 1.15$ at $(i - i^*)/\varepsilon^* = 0.2$, the value of k is found to be 0.75. Thus, at i = 2 deg, the actual fluid deflection is

$$\varepsilon = \varepsilon^* + 0.75\,(i - i^*)$$
$$= 22.24 + 0.75\,(2 - 1.69) = 22.47 \text{ deg}$$

The actual outlet angle is

$$\alpha_2 = \alpha_1 - \varepsilon = \alpha_1' + i - \varepsilon = 50 + 2 - 22.47 = 29.53 \text{ deg}$$

The lift coefficient is determined using eqn. (3.18),

$$C_L = 2(s/l) \cos \alpha_m (\tan \alpha_1 - \tan \alpha_2) - C_D \tan \alpha_m$$

$$\tan \alpha_m = \frac{1}{2}(\tan \alpha_1 + \tan \alpha_2) = \frac{1}{2}(\tan 52^\circ + \tan 29.53^\circ)$$

$$= \frac{1}{2}(1.280 + 0.5665) = 0.9232$$

$$\therefore \alpha_m = 42.71 \text{ deg} \quad \text{and} \quad \cos\alpha_m = 0.7348$$

$$\therefore C_L = 2 \times 0.8 \times 0.7348\,(1.280 - 0.5665) - 0.02 \times 0.9232$$

$$= \underline{0.820}$$

3.6. (a) Show that the pressure rise coefficient $C_p = \Delta p/(\frac{1}{2}\rho c_1^2)$ of a compressor cascade is related to the diffuser efficiency η_D and the total pressure loss coefficient ζ by the following expressions:

$$C_p = \eta_D(1 - \sec^2\alpha_2/\sec^2\alpha_1) = 1 - (\sec^2\alpha_2 + \zeta)/\sec^2\alpha_1$$

where

$$\eta_D = \Delta p/\left(\tfrac{1}{2}\rho(c_1^2 - c_2^2)\right)$$

$$\zeta = \Delta p_o/(\tfrac{1}{2}\rho c_x^2)$$

$$\alpha_1, \alpha_2 = \text{flow angles at cascade inlet and outlet.}$$

(b) Determine a suitable <u>maximum</u> inlet flow angle of a compressor cascade having a space/chord ratio 0.8 and $\alpha_2 = 30$ deg when the diffusion factor D_F is limited to 0.6. The definition of diffusion factor which should be used is the early Lieblein formula,

$$D_F = \left(1 - \frac{\cos\alpha_1}{\cos\alpha_2}\right) + \left(\frac{s}{l}\right)\frac{\cos\alpha_1}{2}(\tan\alpha_1 - \tan\alpha_2).$$

(c) The stagnation pressure loss derived from flow measurements on the above cascade is 149 Pa when the inlet velocity c_1 is 100 m/s at an air density ρ of 1.2 kg/m^3. Determine the values of

(i) pressure rise;

(ii) diffuser efficiency;

(iii) drag and lift coefficients.

<u>Solution</u>. (a) The loss in total pressure across a compressor cascade due to irreversible processes is, for an incompressible flow,

$$\Delta p_o = p_{o1} - p_{o2} = (p_1 - p_2) + \frac{1}{2}\rho(c_1^2 - c_2^2)$$

$$= -\Delta p + \frac{1}{2}\rho c_1^2\left[1 - (c_2/c_1)^2\right]$$

where $\Delta p = p_2 - p_1$, is the static pressure rise across the cascade. With $c_1 \cos \alpha_1 = c_2 \cos \alpha_2 = c_x$ = constant

$$\Delta p_o/(\tfrac{1}{2}\rho c_1^2) = -\Delta p/(\tfrac{1}{2}\rho c_1^2) + (1 - \cos^2 \alpha_1/\cos^2 \alpha_2)$$

$$\therefore C_p = 1 - \Delta p_o/(\tfrac{1}{2}\rho c_1^2) - \cos^2 \alpha_1/\cos^2 \alpha_2$$

$$= 1 - \zeta \cos^2 \alpha_1 - \cos^2 \alpha_1/\cos^2 \alpha_2$$

$$= 1 - (\zeta + \sec^2 \alpha_2)/\sec^2 \alpha_1 \qquad \text{(i)}$$

From the definition of diffuser efficiency

$$\Delta p = \tfrac{1}{2}\rho(c_1^2 - c_2^2)\eta_D$$

$$\therefore C_p = \eta_D(1 - c_2^2/c_1^2) = \eta_D(1 - \sec^2 \alpha_2/\sec^2 \alpha_1) \qquad \text{(ii)}$$

(b) For a compressor cascade of specified geometry the diffusion factor D_F increases rapidly with increasing inlet flow angle as the positive stall "point" is approached. With α_2 = 30 deg, s/ℓ = 0.8 and D_F = 0.6 substituted in the Lieblein formula:-

$$0.6 = 1 - \cos \alpha_1/0.866 + 0.4(\sin \alpha_1 - 0.5774 \cos \alpha_1)$$

Putting $x = \cos \alpha_1$, $(1 - x^2)^{1/2} = \sin \alpha_1$ and rearranging,

$$x(1/0.866 + 0.4 \times 0.5774) = 0.4\left[1 + (1 - x^2)^{1/2}\right]$$

$$\therefore (3.464x - 1)^2 = 1 - x^2$$

$$\therefore 13x^2 - 6.928x + 1 = 1$$

$$\therefore x = \cos \alpha_1 = 6.928/13 = 0.5329$$

Thus, the maximum inlet flow angle (i.e. for positive stall) to give a diffusion factor D_F = 0.6 is

$$\alpha_1 = \underline{57.8 \text{ deg}}$$

(c) With $c_x = c_1 \cos \alpha_1$ = 100 x cos 57.8° = 53.29 m/s, the total pressure loss coefficient is immediately found, i.e.

$$\zeta = \Delta p_o/(\tfrac{1}{2}\rho c_x^2) = 149/(\tfrac{1}{2} \times 1.2 \times 53.29^2) = 0.0875$$

Using eqn. (i),

$$C_p = 1 - (0.0875 + \sec^2 30^o)/\sec^2 57.8^o$$
$$= 1 - (0.0875 + 1.3333) \times 0.5329^2$$
$$= 0.5965$$

The pressure rise is,

$$\Delta p = p_2 - p_1 = \tfrac{1}{2} C_p \rho c_1^2 = \tfrac{1}{2} \times 0.5965 \times 1.2 \times 10^4$$
$$= \underline{3.579 \text{ kPa}}$$

From eqn. (ii) the diffuser efficiency is,

$$\eta_D = C_p/(1 - \cos^2 \alpha_1/\cos^2 \alpha_2)$$
$$= 0.5965/(1 - 0.5329^2/0.866^2) = 0.5965/0.6213$$
$$= \underline{0.96}$$

The drag coefficient is defined, eqns. (3.16b) and (3.17), as

$$C_D = D/(\tfrac{1}{2}\rho c_m^2 \ell) = s \Delta p_o \cos \alpha_m/(\tfrac{1}{2}\rho c_m^2 \ell)$$
$$= \zeta (s/\ell) \cos^3 \alpha_m$$

where $\tan \alpha_m = \tfrac{1}{2}(\tan \alpha_1 + \tan \alpha_2) = \tfrac{1}{2}(\tan 57.8^o + \tan 30^o) = 1.0827$

$$\therefore \alpha_m = 47.27 \text{ deg}$$

$$\therefore C_D = 0.0875 \times 0.8 \times \cos^3 47.27^o$$
$$= \underline{0.0219}$$

The lift coefficient is defined for a compressor cascade, eqn. (3.18), as

$$C_L = 2(s/\ell)\cos \alpha_m(\tan \alpha_1 - \tan \alpha_2) - C_D \tan \alpha_m$$
$$= 2 \times 0.8 \times \cos 47.27^o(\tan 57.8^o - \tan 30^o) - 0.0219 \times 1.0827$$
$$= 1.0972 - 0.0237$$
$$= \underline{1.074}$$

Chapter 4

Axial Flow Turbines

4.1. Show, for an axial flow turbine stage, that the relative stagnation enthalpy across the rotor row does not change. Draw an enthalpy-entropy diagram for the stage labelling all salient points.

Stage reaction for a turbine is defined as the ratio of the static enthalpy drop in the rotor to that in the stage. Derive expressions for the reaction in terms of the flow angles and draw velocity triangles for reactions of zero, 0.5 and 1.0.

Solution. It is assumed that the axial velocity through the stage is constant, i.e. $c_{x1} = c_{x2} = c_{x3} = c_x$, that the absolute velocity at inlet to the stage c_1 equals the absolute velocity at outlet c_3 and that the flow is adiabatic. Referring to the velocity diagram, Fig. 4.1, and enthalpy-entropy diagram, Fig. 4.2, the specific work done by the stage, which causes the specific stagnation enthalpy of the fluid to decrease, eqn. (4.2), is

$$\Delta W = h_{o1} - h_{o3} = U(c_{y2}+c_{y3}) \qquad \text{(i)}$$

As the nozzle flow is adiabatic and the nozzle does no work, then

$$h_{o1} = h_{o2} \qquad \text{(ii)}$$

From the velocity triangles, using the Cosine Rule,

$$w_2^2 = U^2 + c_2^2 - 2Uc_2\cos(\pi/2 - \alpha_2) = U^2 + c_2^2 - 2Uc_{y2} \qquad \text{(iii)}$$

$$w_3^2 = U^2 + c_3^2 - 2Uc_3\cos(\pi/2 + \alpha_3) = U^2 + c_3^2 + 2Uc_{y3} \qquad \text{(iv)}$$

Subtracting eqn. (iii) from eqn. (iv) and re-arranging,

$$U(c_{y2} + c_{y3}) = \frac{1}{2}(c_2^2 - c_3^2 + w_3^2 - w_2^2)$$

Combining eqns. (i), (ii) and (iii), noting that $h_o = h + \frac{1}{2}c^2$,

$$h_{o2} - h_{o3} = (h_2 - h_3) + \frac{1}{2}(c_2^2 - c_3^2) = \frac{1}{2}(c_2^2 - c_3^2 + w_3^2 - w_2^2)$$

$$\therefore\ h_2 + \frac{1}{2}w_2^2 = h_3 + \frac{1}{2}w_3^2 \qquad \text{(vi)}$$

The relative stagnation enthalpy is defined as $h_{orel} = h + \frac{1}{2}w^2$ and eqn. (vi) shows that it is equal at inlet and outlet of the turbine rotor from which it is deduced that it must be constant through the rotor.

The stage reaction, eqn. (4.17), is

$$R = (h_2 - h_3)/(h_1 - h_3)$$

$$= (h_2 - h_3)/(h_{o1} - h_{o3})$$

noting that $c_1 = c_3$ for a "normal" stage. After substituting for $(h_{o1} - h_{o3})$ from eqn. (i) and $(h_2 - h_3)$ from eqn. (vi),

$$R = \frac{w_3^2 - w_2^2}{2U(c_{y2} + c_{y3})} \qquad \text{(vii)}$$

The numerator is factorised as follows,

$$w_3^2 - w_2^2 = (w_{y3}^2 + c_x^2) - (w_{y2}^2 + c_x^2) = (w_{y3} - w_{y2})(w_{y3} + w_{y2})$$

As $w_{y3} + w_{y2} = c_{y3} + c_{y2}$, eqn. (vii) reduces to

$$R = (w_{y3} - w_{y2})/(2U) = (\tan\beta_3 - \tan\beta_2)\, c_x/(2U) \qquad \text{(viii)}$$

Alternatively, with $w_{y3} = U + c_{y3}$ from the velocity diagram,

$$R = (U + c_{y3} - w_{y2})/(2U = \frac{1}{2} + (\tan\alpha_3 - \tan\beta_2)\, c_x/(2U) \qquad \text{(ix)}$$

and, with $w_{y2} = c_{y2} - U$,

$$R = (2U + c_{y3} - c_{y2})/(2U) = 1 + (\tan\alpha_3 - \tan\alpha_2)\, c_x/(2U) \qquad \text{(x)}$$

The velocity and simplified Mollier diagrams for the three reactions $R = 0$, 0.5 and 1.0 for arbitrary but constant values of flow coefficient c_x/U and stage loading factor $(c_{y2} + c_{y3})/U$ are shown below.

(i) $R = 0$, eqn. (viii) gives $\beta_3 = \beta_2$, hence $w_2 = w_3$ and $h_2 = h_3$.

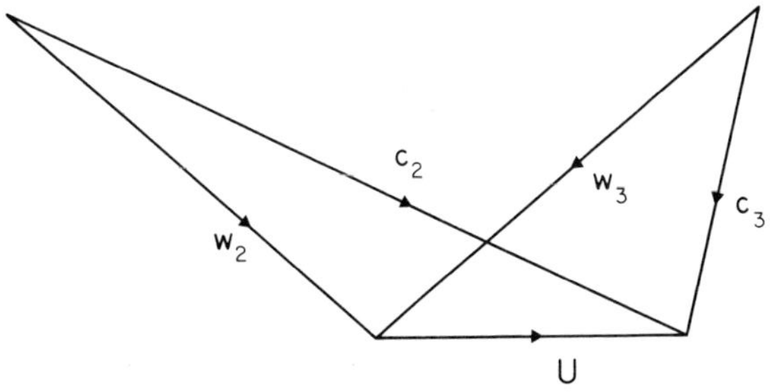

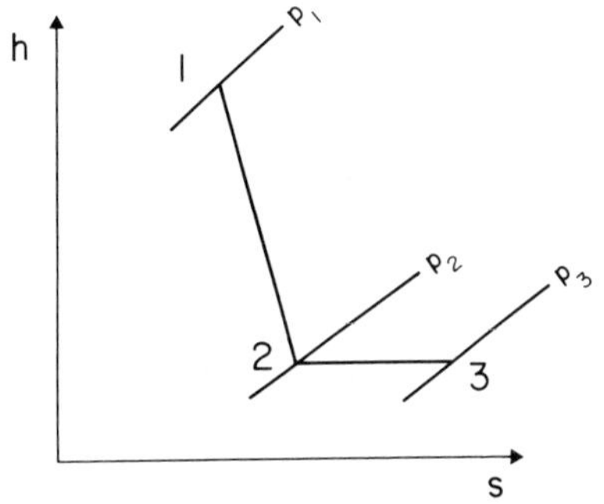

(ii) R = 0.5, eqn. (ix) gives $\alpha_3 = b_2$, hence $c_3 = w_2$, $c_2 = w_3$ and $h_1 - h_2 = h_2 - h_3$.

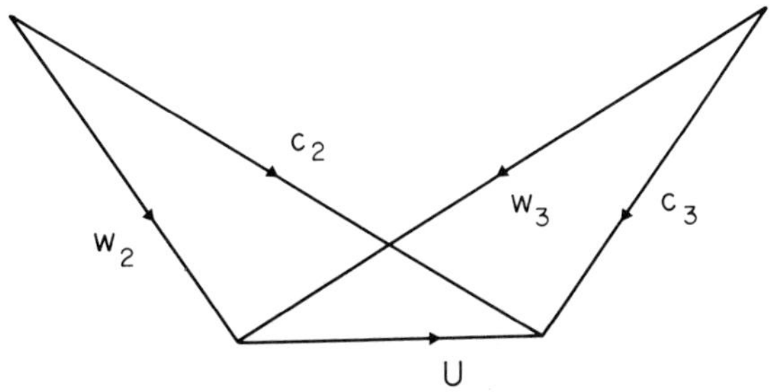

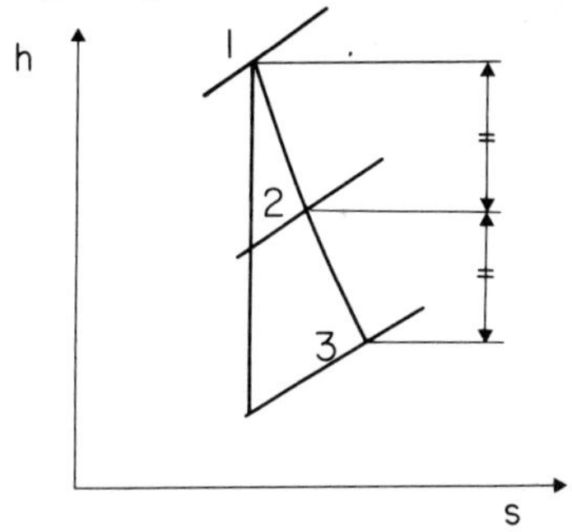

(iii) R = 1.0, eqn. (x) fives $\alpha_3 = \alpha_2$, hence $c_2 = c_3$ and $h_2 = h_1$.

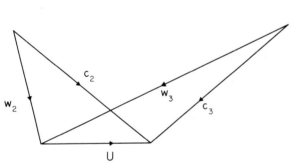

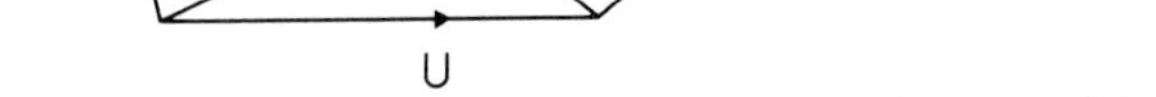

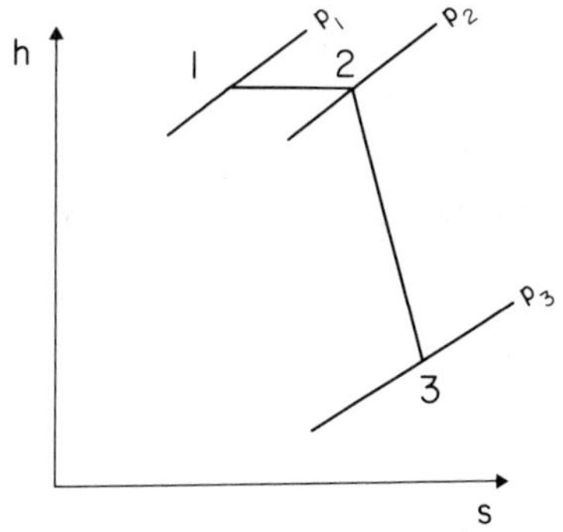

4.2. In a Parsons' reaction turbine the rotor blades are similar to the stator blades but with the angles measured in the opposite direction. The efflux angle relative to each row of blades is 70 deg from the axial direction, the exit velocity of steam from the stator blades is 160 m/s, the blade speed is 152.5 m/s and the axial velocity is constant. Determine the specific work done by the steam per stage.

A turbine of 80% internal efficiency consists of ten such stages as described above and receives steam from the stop valve at 1.5 MPa and 300°C. Determine, with the aid of a Mollier chart, the condition of the steam at outlet from the last stage.

Solution. The velocity diagram for the stage can be readily constructed from the data supplied and the specific work obtained from a scale drawing or, more accurately, by calculation. It will be noticed that as the efflux angle relative to each blade row is equal, i.e. $\alpha_2 = \beta_3 = 70$ deg, the velocity triangles are similar and the reaction is 50 per cent. The specific work per stage is

$$\Delta W = U(c_{y2} + c_{y3})$$

Solving for the unknown swirl velocities using the usual sign convention

$$c_{y2} = c_2 \sin \alpha_2 = 160 \sin 70^\circ$$
$$= 150.4 \text{ m/s}$$
$$c_{y3} = w_3 \sin \beta_3 - U = c_2 \sin \alpha_2 - U$$
$$= 150.4 - 152.5 = -2.1 \text{ m/s}$$
$$\therefore \Delta W = 152.5(150.4 - 2.1)$$
$$= \underline{22.62 \text{ kJ/kg}}$$

This stage is rather lightly loaded and the stage loading factor is

$$\psi = \Delta W/U^2 = (c_{y2} + c_{y3})/U = 148.3/152.5$$
$$= 0.9725$$

A turbine with ten similar stages to the one above will produce a specific work of 226.2 kJ/kg and this is equal to the change in stagnation enthalpy of the steam $h_{oA} - h_{oB}$ between turbine inlet (A) and turbine exhaust (B), i.e.

$$h_{oA} - h_{oB} = 226.2 \text{ kJ/kg}$$

It is implied that the "internal" efficiency is the total to total efficiency, defined as

$$\eta_{tt} = (h_{oA} - h_{oB})/(h_{oA} - h_{oBs})$$
$$\therefore h_{oA} - h_{oBs} = 226.2/0.8 = 282.8 \text{ kJ/kg}$$

From steam tables or Mollier chart at $p_{oA} = 1.5$ MPa (15 bar) and $T_{oA} = 300^\circ$C

$$h_{oA} = 3039 \text{ kJ/kg}$$
$$\therefore h_{oB} = 3039 - 226.2 = 2812.8 \text{ kJ/kg}$$

$\therefore h_{oBs} = 2756.2$ kJ/kg

The less laborious method of determining the exhaust steam condition is by plotting these specific enthalpies on a large scale Mollier chart for steam. From such a plot the exhaust steam condition is

p_{oB} = 420 kPa (4.2 bar);

T_{oB} = $177^{o}C$

i.e. the steam is still superheated at exhaust.

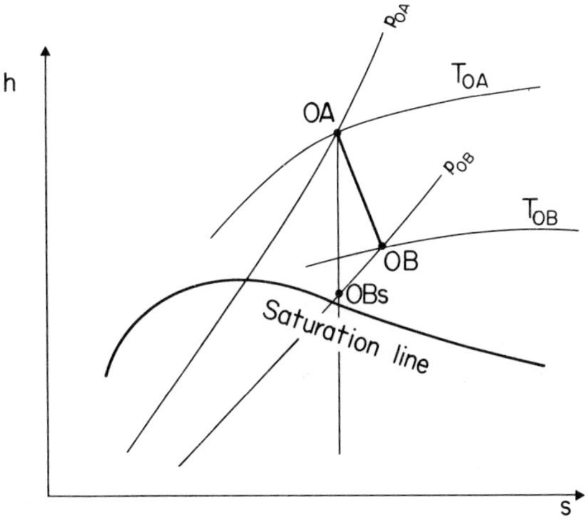

4.3. Values of pressure (kPa) measured at various stations of a zero-reaction gas turbine stage, all at the mean blade height, are shown in the table given below.

Stagnation pressure		Static pressure	
Nozzle entry	414	Nozzle exit	207
Nozzle exit	400	Rotor exit	200

The mean blade speed is 291 m/s, inlet stagnation temperature 1100 K, and the flow angle at nozzle exit is 70° measured from the axial direction. Assuming the magnitude and direction of the velocities at entry and exit of the stage are the same, determine the total to total efficiency of the stage. Assume a perfect gas with $C_p = 1.148$ kJ/(kg ^{o}C) and $\gamma = 1.333$.

Solution. The total to total efficiency of a turbine stage is defined, in the usual notation, as

$$\eta_{tt} = \frac{h_{o1} - h_{o3}}{h_{o1} - h_{o3ss}}$$

With $c_1 = c_3$ this can be rewritten as

$$\eta_{tt} = 1/\left[1 + (h_3 - h_{3ss})/(h_1 - h_3)\right]$$

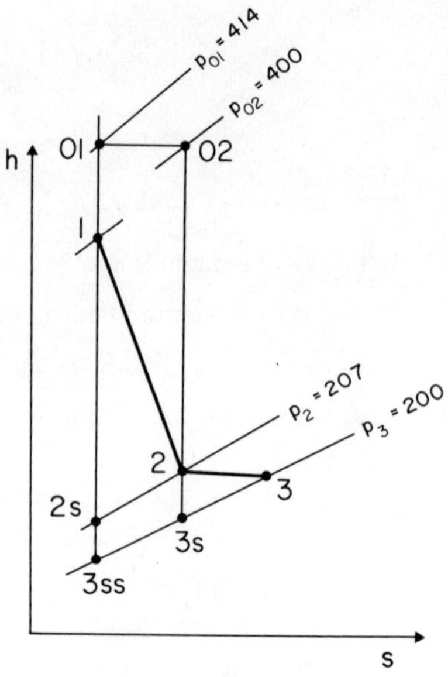

$$\therefore \eta_{tt} = 1/\left[1 + (T_3 - T_{3ss})/(T_1 - T_3)\right]$$

with the perfect gas assumption. In order to determine the efficiency of the stage the velocity diagram must first be solved.

The stage reaction is defined as $R = (h_2 - h_3)/(h_1 - h_3)$ so that zero reaction means h_2 equals h_3. The relative stagnation enthalpy $h_{orel} = h + \frac{1}{2}w^2$ is constant in the rotor, then $h_2 + \frac{1}{2}w_2^2 = h_3 + \frac{1}{2}w_3^2$ and, therefore $w_2 = w_3$. The velocity at nozzle exit c_2 must be determined to complete the velocity diagram. At nozzle exit,

$$T_2 = T_{o2}(p_2/p_{o2})^{(\gamma-1)/\gamma}$$

$$= 1100(207/400)^{0.2498} = 933.1 \text{ K}$$

$$c_2^2 = 2C_p(T_{o2} - T_2) = 2 \times 1148(1100 - 933.1) = 383,200$$

$$\therefore c_2 = 619.1 \text{ m/s}$$

Referring to the velocity diagram, $c_x = c_2 \cos \alpha_2 = 619.1 \cos 70^o = 211.8$ m/s

$$c_{y2} = c_2 \sin \alpha_2 = 619.1 \sin 70^o = 581.8 \text{ m/s}$$

$$\therefore w_{y2} = c_{y2} - U = 581.8 - 291 = 290.8 \text{ m/s}$$

An important point to note is that $w_{y3} = w_{y2}$ $(w_3 = w_2)$. Thus,

$$c_{y3} = w_{y3} - U = w_{y2} - U = 290.8 - 291$$

$$= -0.2 \text{ m/s}$$

i.e. the flow leaving the stage is very nearly axial in direction with a small angle of swirl $\alpha_3 = \tan^{-1}(-0.2/211.8) = -0.05$ deg. Effectively $c_3 = c_1 = c_x = 211.8$ m/s.

Thus, with $T_3 = T_2 = 933.1$ K and $T_{o1} = T_{o2} = 1100$ K

$$T_1 - T_3 = T_{o1} - c_1^2/(2C_p) - T_3$$

$$= 1100 - 211.8^2/(2 \times 1148) - 933.1$$

$$= 147.4^oC$$

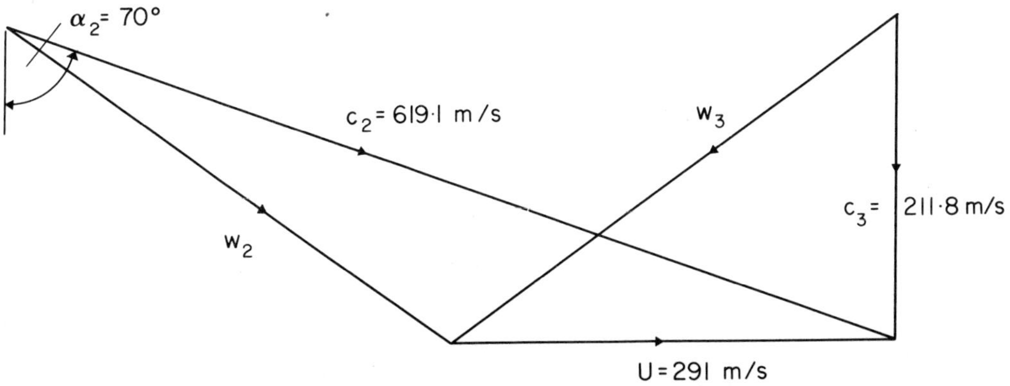

Using the isentropic relation between temperature and pressure

$$T_{3ss} = T_{01}(p_3/p_{01})^{(\gamma-1)/\gamma} = 1100(200/414)^{0.2498} = 917.1\text{ K}$$

$$\therefore T_3 - T_{3ss} = 933.1 - 917.1 = 16.0^{\circ}\text{C}$$

$$\therefore \eta_{tt} = 1/\ 1 + 16/147.4 = \underline{90.2\%}$$

4.4. In a certain axial flow turbine stage the axial velocity c_x is constant. The absolute velocities entering and leaving the stage are in the axial direction. If the flow coefficient c_x/U is 0.6 and the gas leaves the stator blades at 68.2 deg from the axial direction, calculate:

(i) the stage loading factor, $\Delta W/U^2$;

(ii) the flow angles relative to the rotor blades;

(iii) the degree of reaction;

(iv) the total to total and total to static efficiencies.

The Soderberg loss correlation, eqn. (4.12) should be used.

Solution. (i) The stage loading factor is

$$\psi = \Delta W/U^2 = c_{y2}/U, \text{ as } c_{y3} = 0$$

$$= (c_x/U)\tan\alpha_2$$

$$= 0.6 \times \tan 68.2^{\circ} = \underline{1.50}$$

(ii) From the velocity diagram

$$\tan\beta_3 = U/c_x = 1/0.6 = 1.667$$

$$\therefore \underline{\beta_3 = 59.04\text{ deg}}$$

$$\tan \beta_2 = \tan \alpha_2 - U/c_x = 2.5 - 1.667 = 0.8335$$

$$\therefore \beta_2 = 39.81 \text{ deg}$$

(iii) The stage reaction, eqn. (4.22a), is

$$R = (\tan \beta_3 - \tan \beta_2)\, c_x/(2U)$$

$$= 0.3(1.667 - 0.8335) = \underline{0.25}$$

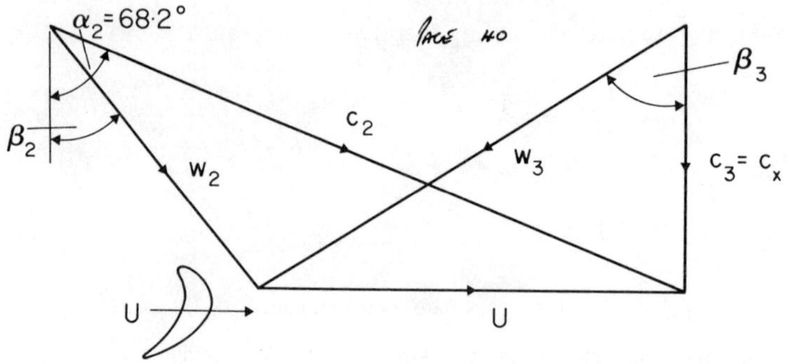

(iv) The total to total efficiency of a normal stage ($c_1 = c_3$) is,

$$\eta_{tt} = (h_{o1} - h_{o3})/(h_{o1} - h_{o3ss}) = (h_1 - h_3)/(h_1 - h_3 + h_3 - h_{3ss})$$

$$= 1/\left[1 + (h_3 - h_{3ss})/(h_1 - h_3)\right]$$

Referring to Fig. 4.2, the enthalpy difference $h_{3s} - h_{3ss}$, equal to $(h_2 - h_{2s})(T_3/T_2)$, is usually simplified to $h_2 - h_{2s}$ with only a small loss in accuracy in determining efficiency.

$$\therefore \eta_{tt} \simeq \left[1 + (h_3 - h_{3s} + h_2 - h_{2s})/(h_1 - h_3)\right]^{-1}$$

The enthalpy differences $h_2 - h_{2s}$ and $h_3 - h_{3s}$ representing the effects of irreversible flow in the nozzle and the rotor respectively, can be expressed in terms of loss coefficients ζ_N and ζ_R,

$$h_2 - h_{2s} = \frac{1}{2} c_2^2 \zeta_N$$

$$h_3 - h_{3s} = \frac{1}{2} w_3^2 \zeta_R$$

Thus, the total to total efficiency becomes, eqn. (4.9a),

$$\eta_{tt} = \left[1 + \frac{\zeta_R w_3^2 + \zeta_N c_2^2}{2(h_1 - h_3)}\right]^{-1} \qquad \text{(i)}$$

The total to static efficiency, defined as

$$\eta_{ts} = (h_{o1} - h_{o3})/(h_{o1} - h_{3ss})$$
$$= 1/\left[1 + (h_3 - h_{3ss} + \tfrac{1}{2}c_1^2)/(h_1 - h_3)\right]$$

is used when the exhaust kinetic energy $\frac{1}{2}c_3^2$ is wasted. This efficiency is most useful in the form,

$$\eta_{ts} = \left[1 + \frac{\zeta_R w_3^2 + \zeta_N c_2^2 + c_1^2}{2(h_1 - h_3)}\right]^{-1} \qquad \text{(ii)}$$

The enthalpy loss coefficients can be expressed, eqn. (4.12), in terms of the fluid deflection ε (deg) of each blade row, that is,

$$\zeta = 0.04\left[1 + 1.5(\varepsilon/100)^2\right]$$

where, for the nozzle, $\varepsilon = \varepsilon_N = \alpha_1 + \alpha_2 = 68.2$ deg (i.e. $\alpha_1 = 0$) and, for the rotor row, $\varepsilon = \varepsilon_R = \beta_2 + \beta_3 = 39.81 + 59.04 = 98.85$ deg. Thus, $\zeta_N = 0.06791$ and $\zeta_R = 0.09863$ after using the above equation.

From eqn. (i), with $w_3 = c_x \sec\beta_3$, $c_2 = c_x \sec\alpha_2$ and $h_1 - h_3 = Uc_x \tan\alpha_2$

$$\eta_{tt} = \left[1 + \frac{\zeta_R \sec^2\beta_3 + \zeta_N \sec^2\alpha_2}{(2\tan\alpha_2)/\phi}\right]^{-1}$$

$$= \left[1 + \frac{0.09863/0.5144^2 + 0.06791/0.3714^2}{2 \times 2.5/0.6}\right]^{-1}$$

$$= \left[1 + \frac{0.865}{8.334}\right]^{-1}$$

$$\therefore \eta_{tt} = \underline{90.6\%}$$

From eqn. (ii), with $c_1 = c_x$

$$\eta_{ts} = \left[1 + (0.865 + 1)/8.334\right]^{-1} = \underline{81.7\%}$$

4.5. A gas turbine stage develops 3.36 MW for a mass flow rate of 27.2 kg/s. The stagnation pressure and stagnation temperature at stage entry are 772 kPa and 1000K. The axial velocity is constant throughout the stage, the gases entering and leaving the stage without any absolute swirl. At nozzle exit the static pressure is 482 kPa and the flow direction is at 18 deg to the plane of the wheel. Determine the axial velocity and degree of reaction for the stage given that the entropy increase in the nozzles is 12.9 J/(kg $^{\circ}$C).

Assume that the specific heat at constant pressure of the gas is 1.148 kJ/(kg $^{\circ}$C) and the gas constant is 0.287 kJ/(kg $^{\circ}$C).

Determine also the total to total efficiency of the stage given that the increase in entropy of the gas across the rotor is 2.7 J/(kg $^{\circ}$C).

Solution. Referring to the Mollier diagram the nozzle exit velocity c_2 is solved by determining T_{2s} from the isentropic temperature-pressure relationship and then estimating the temperature difference $T_2 - T_{2s}$ from the entropy increase across the nozzle. Thus,

$$T_{2s} = T_{o1}(p_2/p_{o1})^{(\gamma-1)/\gamma}$$

$$= 1000(482/772)^{0.25} = 888.9\text{K}$$

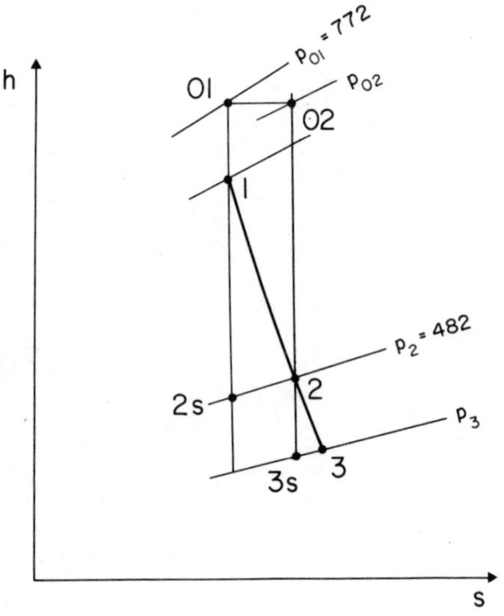

Using the relation $Tds = dh - dp/\rho$, at constant pressure, $T\Delta s \simeq \Delta h$,

$$h_2 - h_{2s} \simeq T_{2s}(s_2 - s_{2s}) = 888.9 \times 12.9$$

$$= 11.47 \text{ kJ/kg}$$

$$\therefore T_2 - T_{2s} = (h_2 - h_{2s})/C_p$$

$$= 11.47/1.148$$

$$= 10^{\circ}\text{C}$$

$$\therefore T_2 = 10 + 888.9 = 898.9\text{K}$$

$$c_2^2 = 2C_p(T_{o1} - T_2) = 2 \times 1148(1000 - 898.9) = 23.21 \times 10^4$$

$$\therefore c_2 = 481.6 \text{ m/s}$$

The axial velocity is easily obtained,

$$c_x = c_2 \cos \alpha_2 = 481.6 \cos(90 - 18)$$

$$= \underline{148.9 \text{ m/s}}$$

The stage reaction is defined, eqn. (4.22c), as

$$R = 1 - (\tan\alpha_2 - \tan\alpha_3) c_x/(2U)$$

which, with $\alpha_3 = 0$, becomes

$$R = 1 - (c_x/2U)\tan\alpha_2$$

The blade speed U is still not determined but can be found from the equation for specific work, viz.,

$$\Delta W = Uc_{y2} = Uc_x \tan\alpha_2$$

$$\therefore U = \Delta W/(c_x \tan\alpha_2) = (\dot{W}/\dot{m})/(c_x \tan\alpha_2)$$

$$= 3.36 \times 10^6/(27.2 \times 148.9 \times \tan 72^\circ) = 269.5 \text{ m/s}$$

$$\therefore R = 1 - 148.9 \times \tan 72^\circ/(2 \times 269.5) = 1 - 0.850$$

$$= \underline{0.150}$$

The total to total efficiency of the stage is

$$\eta_{tt} \doteqdot 1/\left[1 + (h_2 - h_{2s} + h_3 - h_{3s})/(h_{01} - h_{03})\right]$$

$$= 1/\left[1 + C_p(T_2 - T_{2s} + T_3 - T_{3s})/\Delta W\right]$$

The temperature difference $T_3 - T_{3s} \doteqdot T_3(s_3 - s_{3s})/C_p$ and requires the evaluation of T_3. Now

$$\Delta W = h_{01} - h_{03} = h_{01} - h_3 - \tfrac{1}{2}c_x^2, \quad (\text{as } c_3 = c_1 = c_x)$$

$$\therefore T_3 = T_{01} - (\Delta W + \tfrac{1}{2}c_x^2)/C_p$$

$$= 1000 - (123.5 \times 10^3 + \tfrac{1}{2} \times 148.9^2)/1148 = 1000 - 117.2$$

$$= 882.8\text{ K}$$

$$T_3 - T_{3s} \doteqdot 882.8 \times 2.7/1148 = 2.08^\circ\text{C}$$

$$\therefore \eta_{tt} = 1/\left[1 + 1.148(10 + 2.08)/123.5\right] = 1/(1 + 0.1123)$$

$$= \underline{89.9\%}$$

4.6. Derive an approximate expression for the total to total efficiency of a turbine stage in terms of the enthalpy loss coefficients for the stator and rotor when the absolute velocities at inlet and outlet are not equal.

A steam turbine stage of high hub/tip ratio is to receive steam at a stagnation pressure and temperature of 1.5 MPa and 325 $^{\circ}$C respectively. It is designed for a blade speed of 200 m/s and the following blade geometry was selected:

	Nozzles	Rotor
Inlet angle, deg	0	48
Outlet angle, deg	70.0	56.25
Space/chord ratio, s/ℓ	0.42	-
Blade length/axial chord ratio, H/b	2.0	2.1
Max. thickness/blade chord	0.2	0.2

The deviation angle of the flow from the rotor row is known to be 3 deg on the evidence of cascade tests at the design condition. In the absence of cascade data for the nozzle row, the designer estimated the deviation angle from the approximation $0.19\theta s/\ell$ where θ is the blade camber in degrees. Assuming the incidence onto the nozzles is zero, the incidence onto the rotor 1.04 deg and the axial velocity across the stage is constant, determine:

(i) the axial velocity;

(ii) the stage reaction and loading factor;

(iii) the approximate total to total stage efficiency on the basis of Soderberg's loss correlation, assuming Reynolds number effects can be ignored;

(iv) by means of a large steam chart (Mollier diagram) the stagnation temperature and pressure at stage exit.

Solution. The total to total efficiency of a turbine stage, applicable to the case c_1 not equal to c_3, is

$$\eta_{tt} = (h_{o1} - h_{o3})/(h_{o1} - h_{o3ss}) = \Delta W/(\Delta W + \text{"losses"})$$

$$= 1/\left[1 + (h_{o3} - h_{o3ss})/\Delta W\right]$$

Employing the approximations $h_{o3} - h_{o3ss} = h_3 - h_{3ss}$ (i.e. this assumes $c_3 = c_{3ss}$) and $h_{3s} - h_{3ss} = h_2 - h_{2s}$, then

$$\eta_{tt} = 1/\left[1 + (h_2 - h_{2s} + h_3 - h_{3s})/\Delta W\right]$$

Defining the enthalpy loss coefficients, eqns. (4.8a) and (4.8b),

$$h_2 - h_{2s} = \frac{1}{2} c_2^2 \zeta_N$$

for the nozzle (i.e. stator) row, and

$$h_3 - h_{3s} = \frac{1}{2} w_3^2 \zeta_R$$

for the rotor row, the required expression for the efficiency is

$$\eta_{tt} = \left[1 + (\zeta_N c_2^2 + \zeta_R w_3^2)/(2\,\Delta W)\right]^{-1} \qquad \text{(i)}$$

where $\Delta W = h_{01} - h_{03} = U(c_{y2} + c_{y3})$.

(i) The flow directions at inlet and exit of the nozzle and rotor blades are obtained from the blade angles with suitable corrections for the incidence i and deviation δ of each blade row. At nozzle exit the deviation is

$$\delta_N = 0.19\,\theta\, s/\ell$$
$$= 0.19 \times 70 \times 0.42 = 5.6 \text{ deg}$$

Thus, the nozzle exit flow angle is

$$\alpha_2 = \alpha_2' - \delta_N = 70 - 5.6 = 64.4 \text{ deg}$$

For the rotor, the relative flow exit angle is

$$\beta_3 = \beta_3' - \delta_R = 56.25 - 3 = 53.25 \text{ deg}$$

and the relative flow inlet angle is

$$\beta_2 = \beta_2' + i = 48 + 1.04 = 49.04 \text{ deg}$$

From the velocity diagram, $U = c_x(\tan\alpha_2 - \tan\beta_2)$, therefore,

$$c_x = c_1 = U/(\tan\alpha_2 - \tan\beta_2) = 200/(\tan 64.4 - \tan 49.04)$$
$$= \underline{213.9 \text{ m/s}}$$

(ii) The stage reaction is defined, eqn. (4.22a), as

$$R = (c_x/2U)(\tan\beta_3 - \tan\beta_2)$$
$$= (213.9/400)(\tan 53.25 - \tan 49.04)$$
$$= (213.9/400)(1.3392 - 1.1520)$$
$$= \underline{0.10}$$

The stage loading factor is

$$\psi = \Delta W/U^2 = (w_{y2} + w_{y3})/U = (c_x/U)(\tan\beta_2 + \tan\beta_3)$$
$$= (1.152 + 1.3392)\,213.9/200$$
$$= \underline{2.664}$$

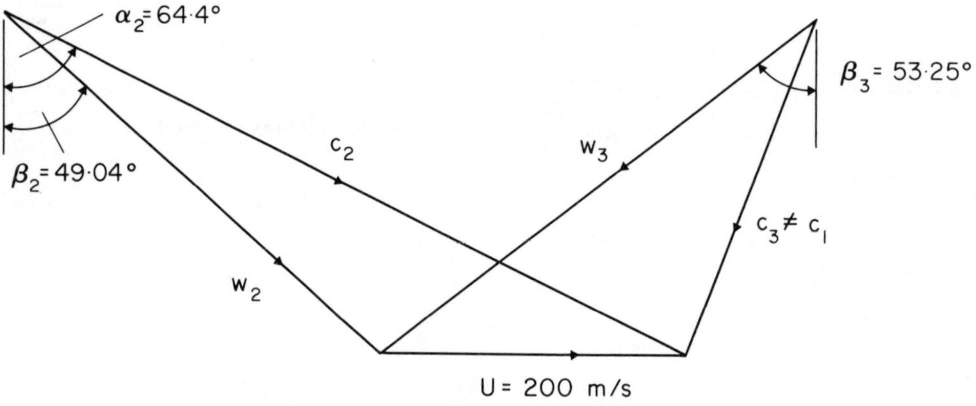

(iii) The total to total efficiency of the stage for the case when c_3 is not equal to c_1 is given by eqn. (i) above. The enthalpy loss coefficients for the nozzle and rotor are evaluated using the analytical simplification of Soderberg's loss correlation, eqn. (4.12), $\zeta^* = 0.04\left[1 + 1.5(\varepsilon/100)^2\right]$ and making suitable corrections for blade aspect ratio in each case.

For the nozzle row, at the nominal aspect ratio H/b = 3.0,

$$\zeta_N^* = 0.04\left[1 + 1.5 \times 0.644^2\right] = 0.06488$$

as the flow deflection in the nozzle row, $\varepsilon_N = \alpha_2$ (i.e. $\alpha_1 = 0$).

At aspect ratios other than the nominal, the enthalpy loss coefficient ζ_{N1} for nozzles can be found, eqn. (4.13a),

$$1 + \zeta_{N1} = (1 + \zeta_N^*)(0.993 + 0.021\,b/H)$$
$$= (1.06488)(0.993 + 0.021/2) = 1.06861$$

$$\therefore \zeta_{N1} = 0.06861$$

For the rotor row, at H/b = 3.0,

$$\zeta_R^* = 0.04\left[1 + 1.5 \times 1.023^2\right] = 0.1028$$

where the flow deflection in the rotor, $\varepsilon_R = \beta_2 + \beta_3 = 102.3$ deg. The correction for the aspect ratio in the case of a rotor row, eqn. (4.13b), is

$$1 + \zeta_{R1} = (1 + \zeta_R^*)(0.975 + 0.075\ b/H)$$

$$= 1.1028(0.975 + 0.075/2.1) = 1.1146$$

$$\therefore \zeta_{R1} = 0.1146$$

The quantities in eqn. (i) are evaluated separately for convenience, i.e.

$$c_2 = c_x \sec \alpha_2 = 213.9 \sec 64.4^\circ = 495.0 \text{ m/s}$$

$$\therefore \zeta_{N1} c_2^2 = 0.06861 \times 495^2 = 16.8 \times 10^3 \text{ m}^2/\text{s}^2 = 16.8 \text{ kJ/kg}$$

$$w_3 = c_x \sec \beta_3 = 213.9 \times \sec 53.25^\circ = 357.5 \text{ m/s}$$

$$\therefore \zeta_{R1} w_3^2 = 0.1146 \times 357.5^2 = 14.65 \times 10^3 \text{ m}^2/\text{s}^2 = 14.65 \text{ kJ/kg}$$

$$\Delta W = \psi U^2 = 2.664 \times 200^2 = 106.6 \text{ kJ/kg}$$

Using eqn. (i),

$$\eta_{tt} = \left[1 + (14.65 + 16.8)/(2 \times 106.6)\right]^{-1} = 1.475^{-1}$$

$$= \underline{87.15\%}$$

(iv) At $p_{o1} = 1.5$ MPa (15 bar), $T_{o1} = 325^\circ$C, the stagnation enthalpy at entry $h_{o1} = 3093.5$ kJ/kg is obtained (tables). Now

$$h_{o1} - h_{o3ss} = (h_{o1} - h_{o3})/\eta_{tt} = \Delta W/\eta_{tt} = 106.6/0.8715$$

$$= 122.3 \text{ kJ/kg}$$

$$\therefore h_{o3ss} = 3093.5 - 122.3 = 2971.2 \text{ kJ/kg}$$

From the Mollier chart, $p_{o3} = \underline{0.9 \text{ MPa }}(9.0 \text{ bar})$.

$$h_{o3} = 3093.5 - 106.6 = 2986.9 \text{ kJ/kg}$$

$$\therefore T_{o3} = \underline{269^\circ\text{C}}$$

Chapter 5
Axial Flow Compressors

Note. In problems 5.1 to 5.4 assume that the gas constant $R = 287\ \text{J/(kg}\ ^{\circ}\text{C)}$ and that $\gamma = 1.4$.

5.1. An axial flow compressor is required to deliver 50 kg/s of air at a stagnation pressure of 500 kPa. At inlet to the first stage the stagnation pressure is 100 kPa and the stagnation temperature is 23°C. The hub and tip diameters at this location are 0.436 m and 0.728 m. At the mean radius, which is constant through all stages of the compressor, the reaction is 0.50 and the absolute air angle at stator exit is 28.8 deg for all stages. The speed of the rotor is 8000 rev/min. Determine the number of similar stages needed assuming that the polytropic efficiency is 0.89 and that the axial velocity at the mean radius is constant through the stages and equal to 1.05 times the average axial velocity.

Solution. The number of stages is determined from the stagnation temperature rise per stage ΔT_o, obtained from the specific work done equation and velocity diagram, and the overall stagnation temperature rise through the compressor, $T_{oB} - T_{oA}$, obtained from the overall stagnation pressure ratio, p_{oB}/p_{oA}, together with the polytropic efficiency, η_p. The number of identical compressor stages, n, is obtained to the nearest integer from

$$n = (T_{oB} - T_{oA})/\Delta T_o \qquad \text{(i)}$$

The specific work done by the rotor on the air, eqn. (5.1), is

$$\Delta W = h_{o2} - h_{o1} = C_p \Delta T_o = U(c_{y2} - c_{y1}) \qquad \text{(ii)}$$

Referring to the mean radius velocity diagram and noticing the velocity triangles are symmetrical for a reaction of 0.5 (i.e. $\beta_2 = \alpha_1$),

$$c_{y2} - c_{y1} = U - 2c_x \tan \alpha_1$$

and, from eqn. (ii)

$$\Delta T_o = U(U - 2c_x \tan \alpha_1)/C_p \qquad \text{(iii)}$$

The average axial velocity $\bar{c}_x$ is obtained from equation of continuity, $\dot{m} = \rho A \bar{c}_x$, the density being determined with the incompressible flow approximation $\rho = \rho_{o1} = p_{o1}/(RT_{o1})$. Thus,

$$\rho_{o1} = p_{o1}/(RT_{o1}) = 10^5/(287 \times 296) = 1.177 \text{ kg/m}^3$$

$$\bar{c}_x = 4\dot{m}/[\rho_{o1} \pi (d_{t1}^2 - d_{h1}^2)]$$

$$= 4 \times 50/[\pi \times 1.177 (0.728^2 - 0.436^2)]$$

$$= 159.1 \text{ m/s}$$

The axial velocity at the mean radius is

$$c_x = 1.05 \times \bar{c}_x = 167.1 \text{ m/s}$$

The mean blade speed is

$$U = \pi N d_m/60 = \pi N(d_{h1} + d_{t1})/120$$

$$= \pi \times 8000(0.436 + 0.728)/120 = 243.8 \text{ m/s}$$

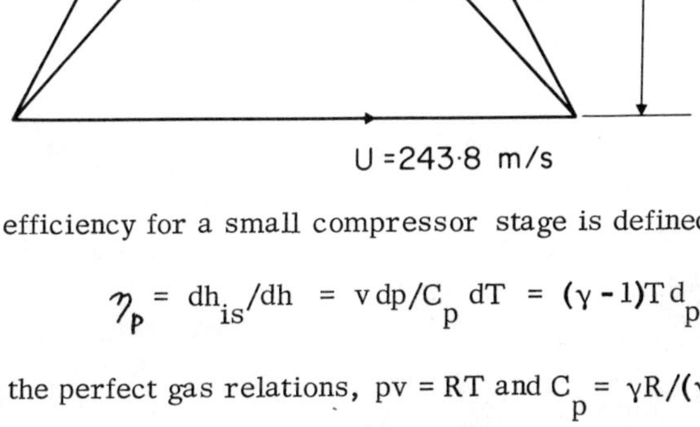

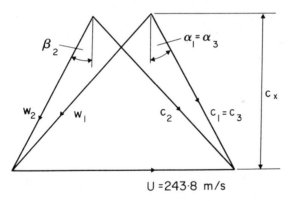

Mean radius velocity diagram

Polytropic efficiency for a small compressor stage is defined, eqn. (2.31), as

$$\eta_p = dh_{is}/dh = v\,dp/C_p\,dT = (\gamma - 1)T\,d_p/(\gamma p\,dT)$$

after using the perfect gas relations, $pv = RT$ and $C_p = \gamma R/(\gamma - 1)$.

$$\therefore T = \text{constant} \times p^{(\gamma-1)/\gamma \eta_p} \qquad \text{(iv)}$$

As the stages are similar with identical velocities, stagnation conditions can be used in eqn. (iv). Thus, across the whole compressor,

$$T_{oB}/T_{oA} = (p_{oB}/p_{oA})^{(\gamma-1)/\gamma\eta_p}$$
$$= 5^{1/(3.5 \times 0.89)}$$
$$= 5^{1/3.115}$$
$$= 1.6764$$
$$\therefore T_{oB} - T_{oA} = 0.6764 \times 296$$
$$= 200.2^{\circ}C$$

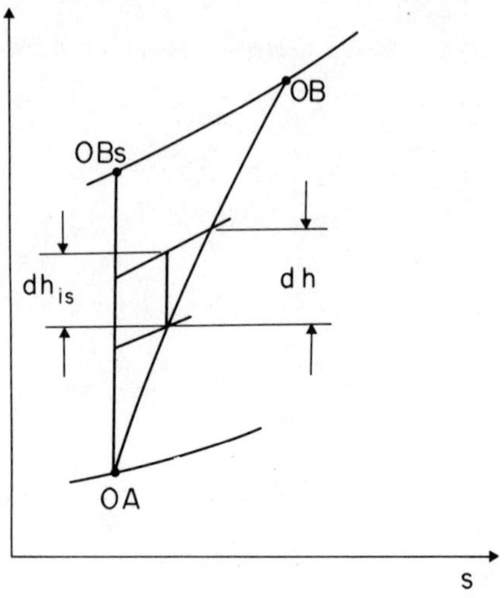

From eqn. (iii),

$$\Delta T_o = 243.8(243.8 - 2 \times 167.1 \times \tan 28.8^{\circ})$$
$$= 14.57^{\circ}C$$

Using eqn. (i)

$$n = 200.2/14.57 = 13.74$$

$\therefore$ The number of stages required is 14.

5.2. Derive an expression for the degree of reaction of an axial compressor stage in terms of the flow angles relative to the rotor and the flow coefficient.

Data obtained from early cascade tests suggested that the limit of efficient working of an axial-flow compressor stage occurred when

(i) a relative Mach number of 0.7 onto the rotor is reached;
(ii) the flow coefficient is 0.5;
(iii) the relative flow angle at rotor outlet is 30 deg measured from the axial direction;
(iv) the stage reaction is 50%.

Find the limiting stagnation temperature rise which would be obtained in the first stage of an axial compressor working under the above conditions and compressing air at an inlet stagnation temperature of 289 K. Assume the axial velocity is constant across the stage.

Solution. The degree of reaction of an axial flow compressor stage is defined as the static enthalpy rise in the rotor divided by the static enthalpy rise in the stage, i.e.

$$R = (h_2 - h_1)/(h_3 - h_1) \qquad \text{(i)}$$

As the relative stagnation enthalpy is constant in the rotor, then

$$h_2 - h_1 = \frac{1}{2}(w_1^2 - w_2^2)$$

Assuming a normal stage (i.e. $c_1 = c_3$), then

$$h_3 - h_1 = h_{o3} - h_{o1} = \Delta W = U(c_{y2} - c_{y1})$$

Substituting into eqn. (i)

$$R = (w_1^2 - w_2^2)/\left[2U(c_{y2} - c_{y1})\right]$$

$$= (w_{y1} + w_{y2})(w_{y1} - w_{y2})/\left[2U(c_{y2} - c_{y1})\right] \qquad \text{(ii)}$$

where it is assumed that c_x is constant across the stage. From the velocity triangles for the compressor stage, $c_{y2} = U - w_{y2}$ and $c_{y1} = U - w_{y1}$ so that $c_{y2} - c_{y1} = w_{y1} - w_{y2}$. Simplifying eqn. (ii),

$$R = (w_{y1} + w_{y2})/(2U) = \phi(\tan\beta_1 + \tan\beta_2)/2 \qquad \text{(iii)}$$

where the flow coefficient $\phi = c_x/U$.

The data given in the problem enables the velocity diagram shape to be drawn

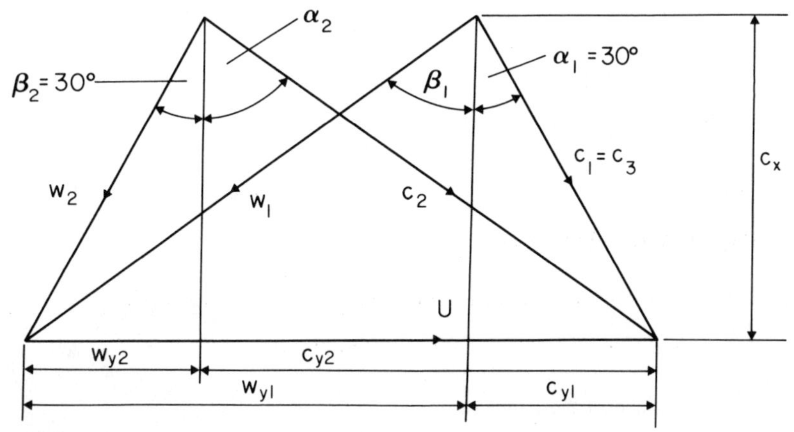

immediately. The magnitudes of the velocity vectors must be calculated from the information concerning maximum relative Mach number. From the velocity diagram the maximum relative velocity is w_1 and the corresponding relative Mach number

$$M_{r1} = w_1/(\gamma R T_1)^{1/2} \qquad \text{(iv)}$$

where the static temperature $T_1 = T_{o1} - c_1^2/(2C_p)$. It is most convenient to solve in terms of the axial velocity c_x. Writing $w_1 = c_x/\cos \beta_1$ and $c_1 = c_x/\cos \alpha_1 = c_x/0.866$, eqn. (iv) gives,

$$w_1^2 = \gamma R M_{r1}^2 T_1$$
$$= \gamma R M_{r1}^2 \left[T_{o1} - c_1^2/(2C_p)\right]$$
$$c_x^2 = \gamma R M_{r1}^2 \left[T_{o1} - c_x^2/(1.5\, C_p)\right] \cos^2 \beta_1 \qquad \text{(v)}$$

Using the equation (iii), β_1 can be determined as follows,

$$\tan \beta_1 = 2R/\phi - \tan \beta_2$$
$$= 2 - \tan 30^{\circ} = 1.4227$$
$$\therefore \beta_1 = 54.9 \text{ deg}$$

Substituting values into eqn. (v),

$$c_x^2 = 1.4 \times 287 \times 0.49\left[289 - c_x^2/(1.5 \times 1005)\right] \times 0.5751^2$$
$$= 1.882 \times 10^4 - 0.0432\, c_x^2$$
$$\therefore c_x = 134.3 \text{ m/s}$$

The stagnation temperature rise in the stage ΔT_o can now be immediately determined using the equation for the specific work,

$$\Delta W = C_p \Delta T_o = U(c_{y2} - c_{y1}) = U(w_{y1} - w_{y2})$$
$$= c_x^2(\tan \beta_1 - \tan \beta_2)/\phi$$
$$\therefore \Delta T_o = c_x^2(\tan \beta_1 - \tan \beta_2)/(\phi C_p)$$
$$= 134.3^2(\tan 54.9^{\circ} - \tan 30^{\circ})/(0.5 \times 1005)$$
$$= \underline{30.35^{\circ}\text{C}}$$

5.3. Each stage of an axial flow compressor is of 0.5 reaction, has the same mean blade speed and the same flow outlet angle of 30 deg relative to the blades. The mean flow coefficient is constant for all stages at 0.5. At entry to the first stage the stagnation temperature is 278 K, the stagnation pressure 101.3 kPa, the static pressure is 87.3 kPa and the flow area 0.372 m^2. Using compressible flow analysis determine the axial velocity and the mass flow rate.

Determine also the shaft power needed to drive the compressor when there are 6 stages and the mechanical efficiency is 0.99.

Solution. It is tactitly assumed that the flow preceding the first stage is deflected by inlet guide vanes to give an absolute flow angle α_1 of 30 deg, the same as all the other stages. The absolute inlet flow velocity c_1 is determined from the stagnation enthalpy definition

$$h_{o1} = h_1 + \frac{1}{2}c_1^2$$

$$\therefore c_1^2 = 2C_p(T_{o1} - T_1)$$

where $C_p = \gamma R/(\gamma - 1)$, and the isentropic temperature-pressure relationship,

$$T_1/T_{o1} = (p_1/p_{o1})^{(\gamma-1)/\gamma}$$

$$= (87.3/101.3)^{1/3.5} = 0.9584$$

$$\therefore c_1^2 = 2C_pT_{o1}(1 - T_1/T_{o1}) = 2 \times 1005 \times 278(1 - 0.9584)$$

$$= 2.325 \times 10^4$$

$$c_1 = 152.5 \text{ m/s}$$

Thus, the axial velocity is

$$c_x = c_1 \cos \alpha_1 = 152.5 \cos 30^o$$

$$= \underline{132.1 \text{ m/s}}$$

Using the equation of continuity, the mass flow rate is

$$\dot{m} = \rho_1 A_1 c_x$$

where

$$\rho_1 = p_1/(RT_1) = 87.3 \times 10^3/(287 \times 0.9584 \times 278)$$

$$= 1.1417 \text{ kg/m}^3$$

$$\therefore \dot{m} = 1.1417 \times 0.372 \times 132.1$$

$$= \underline{56.1 \text{ kg/s}}$$

The specific work done on the gas per stage is

$$\Delta W = U(c_{y2} - c_{y1}) = U(U - 2c_x \tan \alpha_1)$$
$$= U^2(1 - 2\phi \tan \alpha_1)$$

as the velocity triangles are similar for a reaction of 0.5.

$$\therefore \Delta W = (2 \times 132.1)^2 (1 - \tan 30^\circ)$$
$$= 29.5 \text{ kJ/kg}$$

The shaft power needed to drive the compressor (including mechanical losses) is

$$\dot{W}_c = n\dot{m} \, \Delta W/\eta_m$$

where n is the number of stages and η_m the mechanical efficiency. Thus,

$$\dot{W}_c = 6 \times 56.1 \times 29.5 \times 10^3/0.99$$
$$= \underline{10.03 \text{ MW}}$$

5.4. A sixteen-stage axial flow compressor is to have a pressure ratio of 6.3. Tests have shown that a stage total to total efficiency of 0.9 can be obtained for each of the first six stages and 0.89 for each of the remaining ten stages. Assuming constant work done in each stage and similar stages find the compressor overall total to total efficiency. For a mass flow rate of 40 kg/s determine the power required by the compressor. Assume an inlet total temperature of 288 K.

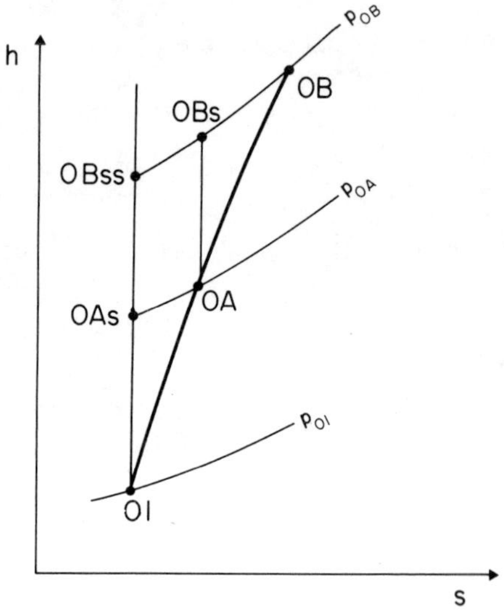

Solution. The sketch shows the overall compressor process in the form of a Mollier diagram. The overall stagnation pressure ratio, p_{oB}/p_{o1}, is the product of the pressure ratio for the first six stages, p_{oA}/p_{o1}, and the pressure ratio for the remaining ten stages, p_{oB}/p_{oA}. i.e.

$$p_{oB}/p_{o1} = (p_{oA}/p_{o1})(p_{oB}/p_{oA}) = 6.3 \quad \text{(i)}$$

It is convenient to assume that the respective stage efficiencies of these two groups of stages is equal to the polytropic (i.e. infinitesimally small stage) efficiency of each group. Using eqn. (2.33), the stagnation pressure ratios in terms of the actual stagnation temperatures and polytropic efficiencies are

$$p_{oA}/p_{o1} = (T_{oA}/T_{o1})^{\gamma\eta_p/(\gamma-1)} = (T_{oA}/T_{o1})^{3.15} \qquad \text{(ii)}$$

with $\eta_p = 0.9$ and $\gamma = 1.4$, and

$$p_{oB}/p_{oA} = (T_{oB}/T_{oA})^{\gamma\eta_p/(\gamma-1)} = (T_{oB}/T_{oA})^{3.115} \qquad \text{(iii)}$$

with $\eta_p = 0.89$ and $\gamma = 1.4$.

The work done in each stage is assumed to be constant so that the stagnation temperature rise for each stage ΔT_o is constant (C_p is constant). For convenience put $x = \Delta T_o/T_{o1}$. The two temperature ratios can now be written as

$$T_{oA}/T_{o1} = 6\Delta T_o/T_{o1} + 1 = 1 + 6x$$

$$T_{oB}/T_{oA} = 10\Delta T_o/T_{oA} + 1 = 1 + 10x/(1+6x)$$

Substituting eqns. (ii) and (iii) into eqn. (i) and using the above temperature ratios,

$$\begin{aligned} 6.3 &= (1+6x)^{3.15} . \left[1 + 10x/(1+6x)\right]^{3.115} \\ &= (1+6x)^{3.15} . \left[(1+16x)/(1+6x)\right]^{3.115} \\ &= (1+16x)^{3.115} . (1+6x)^{0.035} \end{aligned}$$

The unknown x cannot be solved explicitly but can be determined quite easily by a process of trial and error, i.e.

x	0.049	0.050	0.051
$(1+16x)^{3.115}$	6.0687	6.2398	6.4142
$(1+6x)^{0.035}$	1.0091	1.0092	1.0094
RHS	6.1237	6.2974	6.4744

By plotting these values the correct value of x = 0.05002 corresponding to a pressure ratio of 6.3.

Referring again to the Mollier diagram, the overall total to total efficiency of the compressor is

$$\eta_{tt} = (T_{oBss} - T_{o1})/(T_{oB} - T_{o1})$$
$$= \left[(p_{oB}/p_{o1})^{(\gamma-1)/\gamma} - 1\right] T_{o1}/(T_{oB} - T_{o1})$$

where $T_{oB} - T_{o1} = 16\,\Delta T_o = 16x\ T_{o1}$

$$\therefore \eta_{tt} = \left[(p_{oB}/p_{o1})^{(\gamma-1)/\gamma} - 1\right]/(16x)$$
$$= \left[6.3^{1/3.5} - 1\right]/(16 \times 0.05002)$$
$$= 0.6919/0.8003 = \underline{86.45\%}$$

The power required by the compressor (excluding mechanical losses) is given by,

$$\dot{W}_c = \dot{m}\, C_p (T_{oB} - T_{o1})$$
$$= \dot{m}\, C_p\, n\, \Delta T_o = \dot{m}\, C_p\, n\, x\, T_{o1}$$
$$= 40 \times 1005 \times 16 \times 0.05002 \times 288$$
$$= \underline{9.266\ \text{MW}}$$

5.5. At a particular operating condition an axial flow compressor has a reaction of 0.6, a flow coefficient of 0.5 and a stage loading, defined as $\Delta h_o/U^2$ of 0.35. If the flow exit angles for each blade row may be assumed to remain unchanged when the mass flow is throttled, determine the reaction of the stage and the stage loading when the air flow is reduced by 10% at constant blade speed. Sketch the velocity triangles for the two conditions.

Comment upon the likely behaviour of the flow when further reductions in air mass flow are made.

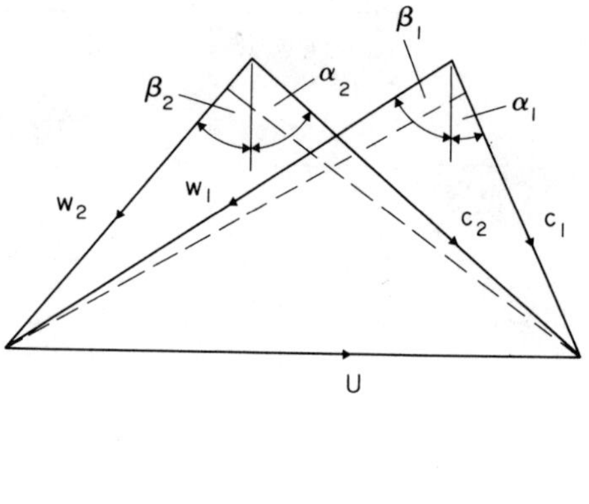

Solution. The velocity diagram shows the velocity vectors for the two conditions (the broken lines denoting the reduced mass flow condition). It is important to notice that it is the flow angle relative to each blade row of the stage which is assumed to remain the same. Cascade test measurements of compressor blade rows indicate that this assumption is only approximately correct for lightly loaded blades with a small space/chord ratio.

The stage loading factor is

$$\psi = \Delta h_o/U^2 = \Delta W/U^2 = (c_{y2} - c_{y1})/U = (w_{y1} - w_{y2})/U$$

$$= \phi(\tan\beta_1 - \tan\beta_2) \qquad \text{(i)}$$

The stage reaction ratio is defined, eqn. (5.11), as

$$R = \phi \tan\beta_m = \phi(\tan\beta_1 + \tan\beta_2)/2 \qquad \text{(ii)}$$

Solving eqns. (i) and (ii) for $\tan\beta_1$ and $\tan\beta_2$

$$\tan\beta_1 = (R + \psi/2)/\phi$$

$$\tan\beta_2 = (R - \psi/2)/\phi$$

At the initial flow coefficient, $\phi = 0.5$, and with $R = 0.6$, $\psi = 0.35$, the relative flow angles are

$$\beta_1 = \tan^{-1}\left[(0.6 + 0.175)/0.5\right] = 57.17 \text{ deg}$$

$$\beta_2 = \tan^{-1}\left[(0.6 - 0.175)/0.5\right] = 40.36 \text{ deg}$$

From the velocity triangles

$$\tan\alpha_1 = 1/\phi - \tan\beta_1 \qquad \text{(iii)}$$

$$\tan\alpha_2 = 1/\phi - \tan\beta_2 \qquad \text{(iv)}$$

and, at the initial flow coefficient, the absolute flow angles are

$$\alpha_1 = \tan^{-1}\left[2 - 1.55\right] = 24.22 \text{ deg}$$

$$\alpha_2 = \tan^{-1}\left[2 - 0.85\right] = 49.0 \text{ deg}$$

The flow angles α_1 and β_2 are assumed to be constant with variation of the flow coefficient. Setting the stage loading factor and reaction in terms of these fixed angles, eqn. (i) can be rewritten

$$\psi = 1 - \phi(\tan\alpha_1 + \tan\beta_2) = 1 - \phi(0.45 + 0.85)$$

$$= 1 - 1.3\,\phi$$

and eqn. (ii) becomes

$$R = 0.5 - \phi(\tan\alpha_1 - \tan\beta_2)/2 = 0.5(1 + 0.4\phi)$$

after using eqns. (iii) and (iv). Thus, at the reduced flow rate, $\phi = 0.9 \times 0.5 = 0.45$ and

$$\underline{R = 0.59}, \quad \underline{\psi = 0.415}$$

Further reduction of the flow rate would eventually lead to the condition of blade stall, probably in the rotor row as this is more heavily loaded than the stator (i.e. $R > 0.5$), with a rapid increase in total pressure losses as stall is approached. The assumption regarding constant flow angle relative to each blade row would also fail to hold as stall implies flow separation off the blade suction surface.

5.6. The proposed design of a compressor rotor blade row is for 59 blades with a circular arc camber line. At the mean radius of 0.254 m the blades are specified with a camber of 30 deg, a stagger of 40 deg and a chord length of 30 mm. Determine, using Howell's correlation method, the nominal outlet angle, the nominal deviation and the nominal inlet angle. The tangent difference approximation, proposed by Howell for nominal conditions ($0 \leqslant \alpha_2^* \leqslant 40^\circ$), can be used:

$$\tan\alpha_1^* - \tan\alpha_2^* = 1.55/(1 + 1.5\, s/\ell).$$

Determine the nominal lift coefficient given that the blade drag coefficient $C_D = 0.017$. Using the data for relative deflection given in Fig. 3.17 determine the flow outlet angle and lift coefficient when the incidence $i = 1.8$ deg. Assume that the drag coefficient is unchanged from the previous value.

Solution. The nominal deviation δ^* can be determined directly from the blade geometry and space/chord ratio s/ℓ using eqns. (3.39) and (3.40), viz.,

$$\delta^* = m\theta(s/\ell)^{1/2}$$

where $\theta = \alpha_1' - \alpha_2'$ is the blade camber, $m = 0.23 + \alpha_2^*/500$, α_1', α_2' are the blade inlet and outlet angles respectively and $\alpha_2^* = \alpha_2' + \delta^*$ is the nominal flow outlet angle.

For blades with a circular arc camber line the blade angles are

$$\alpha_1' = \xi + \theta/2 = 40 + 30/2 = 55 \text{ deg}$$

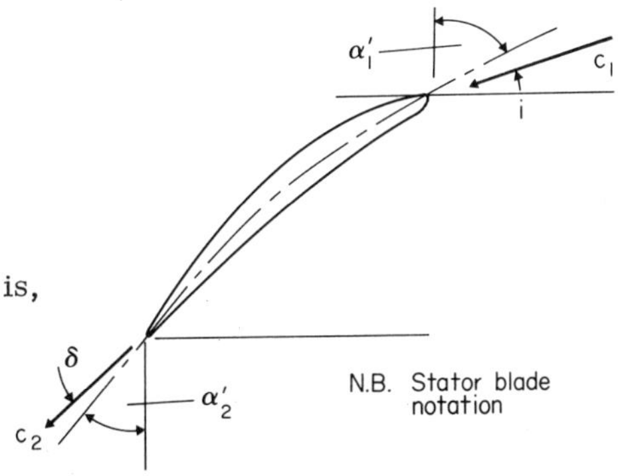

$$\alpha_2 = \xi - \theta/2 = 40 - 30/2 = 25 \text{ deg}$$

where ξ is the stagger angle.

The space/chord ratio at the mean radius is,

$$s/\ell = 2\pi r/(Z\ell)$$
$$= 2\pi \times 0.254/(59 \times 0.03)$$
$$= 0.9017$$

Hence, using Howell's correlation, the nominal flow outlet angle is

$$\alpha_2^* = \alpha_2' + \delta^* = 25 + (0.23 + \alpha_2^*/500)\,30\,(0.9017)^{1/2}$$
$$= 25 + 6.552 + 0.057\,\alpha_2^*$$
$$\therefore \alpha_2^* = 31.55/0.943 = \underline{33.46 \text{ deg}}$$
$$\delta^* = \alpha_2^* - \alpha_2' = 33.46 - 25$$
$$= \underline{8.46 \text{ deg}}$$

Using Howell's tangent difference approximation

$$\tan \alpha_1^* = \tan \alpha_2^* + 1.55/(1 + 1.5\, s/\ell)$$
$$= 0.6609 + 1.55/(1 + 1.5 \times 0.9017)$$
$$= 1.320$$
$$\underline{\alpha_1^* = 52.85 \text{ deg}}$$

Thus, the nominal incidence and nominal deflection angles are

$$i^* = \alpha_1^* - \alpha_1' = 52.85 - 55 = -2.15 \text{ deg}$$
$$\varepsilon^* = \alpha_1^* - \alpha_2^* = 52.85 - 33.46 = 19.39 \text{ deg}$$

From eqn. (3.18), the lift coefficient at the nominal condition is

$$C_L = 2(s/\ell)\cos \alpha_m^* (\tan \alpha_1^* - \tan \alpha_2^*) - C_D \tan \alpha_m^*$$
$$\tan \alpha_m^* = (\tan \alpha_1^* + \tan \alpha_2^*)/2 = (1.32 + 0.6609)/2 = 0.9905$$
$$\alpha_m^* = 44.73 \text{ deg}$$
$$\cos \alpha_m^* = 0.7105$$

$$\therefore C_L = 2 \times 0.9017 \times 0.7105 \times 0.659 - 0.017 \times 0.9905$$

$$= 0.8444 - 0.0168 = \underline{0.8276}$$

In Fig. 3.17 (see sketch) the relative deflection $\varepsilon/\varepsilon^*$ is expressed as a function of the relative incidence $(i - i^*)/\varepsilon^*$. Using this curve the deflection ε at any arbitrary incidence i (within certain limits) can be found provided i^* and ε^* are known. At the given incidence i = 1.8 deg,

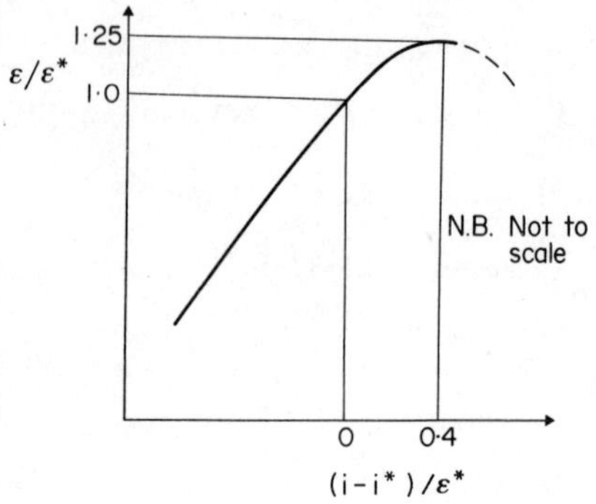

$$(i - i^*)/\varepsilon^* = (1.8 + 2.15)/19.39 = 0.2037$$

$$\therefore \varepsilon/\varepsilon^* = 1.15 \text{ so that } \varepsilon = \alpha_1 - \alpha_2 = 22.3 \text{ deg}$$

Hence, the flow directions are

$$\alpha_1 = i + \alpha_1' = 1.8 + 55 = 56.8 \text{ deg}$$

$$\alpha_2 = \alpha_1 - \varepsilon = 56.8 - 22.3 = \underline{34.5 \text{ deg}}$$

and the deviation angle

$$\delta = \alpha_2 - \alpha_2' = 34.5 - 25 = 9.5 \text{ deg (c.f. } \delta^* = 8.46 \text{ deg)}$$

The lift coefficient at i = 1.8 deg can now be calculated with $\tan \alpha_1 = 1.5282$, $\tan \alpha_2 = 0.6873$, $\tan \alpha_m = 1.1077$, $\alpha_m = 47.93$ deg and $\cos \alpha_m = 0.6701$.

$$\therefore C_L = 2 \times 0.901 \times 0.6701 \times 0.8409 - 0.017 \times 1.1077$$

$$= \underline{0.9974}$$

5.7. The preliminary design of an axial flow compressor is to be based upon a simplified consideration of the mean diameter conditions. Suppose that the stage characteristics of a repeating stage of such a design are as follows:

Stagnation temperature rise	25°C
Reaction ratio	0.6
Flow coefficient	0.5
Blade speed	275 m/s

The gas compressed is air with a specific heat at constant pressure of 1.005 kJ/

(kg ^{o}C). Assuming constant axial velocity across the stage and equal absolute velocities at inlet and outlet, determine the relative flow angles for the rotor.

Physical limitations for this compressor dictate that the space/chord ratio is unity at the mean diameter. Using Howell's correlation method, determine a suitable camber at the mid-height of the rotor blades given that the incidence angle is zero. Use the tangent difference approximation

$$\tan \beta_1^* - \tan \beta_2^* = 1.55/(1 + 1.5\, s/\ell)$$

for nominal conditions and the data of Fig. 3.17 for finding the design deflection. (Hint. Use several trial values of θ to complete the solution.)

Solution. It is usually most convenient to solve the rotor relative flow angles β_1 and β_2 in terms of the stage loading factor ψ, the stage reaction R and the flow coefficient ϕ. The stage loading factor is defined, eqn. (5.14a), as

$$\psi = \Delta W/U^2 = C_p(T_{o3} - T_{o1})/U^2$$

$$= 1005 \times 25/275^2 = 0.3322 \qquad \text{(i)}$$

Referring to the velocity diagram, Fig. 5.2 (or see earlier solutions), and with $\Delta W = U(c_{y2} - c_{y1}) = U(w_{y1} - w_{y2})$,

$$\psi = (w_{y1} - w_{y2})/U = \phi(\tan \beta_1 - \tan \beta_2) \qquad \text{(ia)}$$

where $\phi = c_x/U$.

The reaction of a compressor stage, eqn. (5.11), is

$$R = \phi \tan \beta_m = \phi(\tan \beta_1 + \tan \beta_2)/2 \qquad \text{(ii)}$$

Combining eqns. (ia) and (ii)

$$\tan \beta_1 = (R + \psi/2)/\phi = (0.6 + 0.3322/2)/0.5 = 1.532$$

$$\tan \beta_2 = (R - \psi/2)/\phi = (0.6 - 0.3322/2)/0.5 = 0.8678$$

Thus,

$$\underline{\beta_1 = 56.87 \text{ deg}}, \quad \underline{\beta_2 = 40.95 \text{ deg}}$$

At zero incidence the blade inlet angle $\beta_1' = \beta_1 = 56.87^{o}$. The blade camber

$\theta = \beta_1' - \beta_2' = \beta_1 - (\beta_2+\delta) = \varepsilon + \delta$, where $\varepsilon = \beta_1 - \beta_2 = 15.92$ deg is the fluid deflection and δ is the deviation angle. Thus, as $\delta > 0$, then $\theta > \varepsilon = 15.92$ deg. A suitable value for the camber angle θ may be obtained by a process of trial and error in which several values of θ are selected and corresponding values of ε are determined. Values of ε can be found from Howell's curve of relative deflection $\varepsilon/\varepsilon^*$ against relative incidence $(i - i^*)/\varepsilon^*$, Fig. 3.17, and this means that values of ε^* and i^* must be estimated for each value of θ used.

To simplify the calculation procedure values of the nominal deviation have been estimated from Howell's expression $\delta^* = m\theta(s/\ell)^{1/2}$ with m = constant = 0.26 rather than the more complicated form of m, eqn. (3.40a), used in the solution of the previous problem. Noting that s/ℓ is unity,

$$\varepsilon^* = \beta_1^* - \beta_2^* = \beta_1' - (\beta_2' + \delta^*) = \theta - \delta^* = \theta - 0.26\theta$$

$$= 0.74\theta$$

$$\therefore \beta_2^* = \beta_1^* - \varepsilon^* = \beta_1' - 0.74\theta = 56.87 - 0.74\theta$$

Using the tangent difference approximation,

$$\tan\beta_1^* = \tan\beta_2^* + 1.55/2.5 = \tan(56.87 - 0.74\theta) + 0.62.$$

A tabular form of solution is desirable as follows:

θ^o	20	22.5	25
$\delta^* = 0.26\,\theta$	5.2	5.85	6.5
$\beta_2' = \beta_1' - \theta$	36.87	34.37	31.37
$\beta_2^* = \beta_2' + \delta^*$	42.07	40.22	38.37
$\tan\beta_2^*$	0.9026	0.8457	0.7917
$\tan\beta_1^*$	1.5226	1.4657	1.4117
β_1^*	56.7	55.7	54.69
$i^* = \beta_1^* - \beta_1'$	-0.166	-1.174	-2.182
$\varepsilon^* = \beta_1^* - \beta_2^*$	14.63	15.48	16.32
$(i - i^*)/\varepsilon^*$	0.0113	0.0758	0.1337
$\varepsilon/\varepsilon^*$ (graph)	1.009	1.060	1.105

θ	20	22.5	25
ε	14.76	16.4	18.03

By plotting values of θ against ε (or by numerical interpolation), at ε = 15.92 deg the camber is

$$\underline{\theta = 21.76 \text{ deg}}$$

It is worth commenting that Howell recommended θ to lie in the range

$$1.2\varepsilon^* < \theta < 1.8\varepsilon^*$$

which for $\varepsilon^* \simeq$ 15 deg gives

$$18^\circ < \theta < 27^\circ$$

i.e. the value of θ determined is satisfactory.

Chapter 6

Three-dimensional Flows in Axial Turbomachines

6.1. Derive the radial equilibrium equation for an incompressible fluid flowing with axisymmetric swirl through an annular duct.

Air leaves the inlet guide vanes of an axial flow compressor in radial equilibrium and with a free-vortex tangential velocity distribution. The absolute static pressure and static temperature at the hub, radius 0.3 m, are 94.5 kPa and 293 K respectively. At the casing, radius 0.4 m, the absolute static pressure is 96.5 kPa. Calculate the flow angles at exit from the vanes at the hub and casing when the inlet absolute stagnation pressure is 101.3 kPa. Assume the fluid to be inviscid and incompressible. (Take R = 0.287 kJ/(kg $^{\circ}$C) for air.)

Solution. A detailed derivation of the radial equilibrium equation is presented in FMTT2 and so only a brief outline of the important equations (for an incompressible flow) is given here.

For a fluid element which is in radial equilibrium ($c_r = 0$), rotating about an axis at radius r with a tangential velocity component c_θ, the static pressure gradient is

$$\frac{dp}{dr} = \rho \frac{c_\theta^2}{r} \qquad \text{(i)}$$

The total pressure p_o in an incompressible fluid flow is

$$p_o = p + \frac{1}{2}\rho c^2 = p + \frac{1}{2}\rho(c_x^2 + c_\theta^2) \qquad \text{(ii)}$$

where c_x is the axial component of the velocity c. Differentiating eqn. (ii) with respect to r and combining the result with eqn. (i), the required form of the radial equilibrium equation is found, viz.,

$$\frac{1}{\rho}\frac{dp_o}{dr} = c_x\frac{dc_x}{dr} + c_\theta\frac{dc_\theta}{dr} + \frac{c_\theta^2}{r} = c_x\frac{dc_x}{dr} + \frac{c_\theta}{r}\frac{d}{dr}(rc_\theta) \qquad \text{(iii)}$$

The inlet guide vanes of the axial flow compressor deflect the incoming axial flow away from the meridional plane imparting a free-vortex swirl to the flow. For a free-vortex, $rc_\theta = K = \text{constant}$. Substituting for c_θ in eqn. (i),

$$\frac{1}{\rho}\frac{dp}{dr} = \frac{K^2}{r^3} \qquad \text{(iv)}$$

After integrating eqn. (iv) and putting limits at the hub and tip,

$$\frac{1}{\rho}(p_t - p_h) = \frac{K^2}{2}\left(\frac{1}{r_h^2} - \frac{1}{r_t^2}\right) \qquad \text{(v)}$$

The boundary values given are that at

$$r = r_t = 0.4 \text{ m}, \quad p = p_t = 96.5 \text{ kPa, and at}$$

$$r = r_h = 0.3 \text{ m}, \quad p = p_h = 94.5 \text{ kPa}, \quad T = T_h = 293 \text{ K}.$$

Solving for ρ and K

$$\rho = p_h/(RT_h) = 94.5 \times 10^3/(287 \times 293) = 1.124 \text{ kg/m}^3$$

$$K^2 = \frac{2}{\rho}\frac{(p_t - p_h)}{(1/r_h^2 - 1/r_t^2)} = \frac{2}{1.124}\frac{(96.5 - 94.5) \times 10^3}{(1/0.3^2 - 1/0.4^2)} = 732.1$$

$$\therefore K = rc_\theta = 27.06 \text{ m}^2/\text{s}$$

Thus, for any radius r the magnitude of c_θ can be found. To determine the flow angles, c_x is needed and this can be found with eqn. (ii),

$$c_x^2 = 2(p_o - p)/\rho - (K/r)^2$$

$$= 2(101.3 - 94.5) \times 10^3/1.124 - 732.1/0.3^2 = 3965$$

$$\therefore c_x = 62.97 \text{ m/s} = \text{constant for all radii}$$

$$\tan \alpha_h = \frac{c_{\theta h}}{c_x} = \frac{27.06}{62.97 \times 0.3} = 1.432$$

$$\tan \alpha_t = \frac{c_{\theta t}}{c_x} = \frac{27.06}{62.97 \times 0.4} = 1.074$$

Thus, the flow angles at the hub and tip are, respectively,

$$\underline{\alpha_h = 55.08^\circ}, \quad \underline{\alpha_t = 47.05^\circ}$$

6.2. A gas turbine stage has an initial absolute pressure of 350 kPa and a temperature of 565°C with negligible initial velocity. At the mean radius, 0.36 m, conditions are as follows:

Nozzle exit flow angle	68 deg
Nozzle exit absolute static pressure	207 kPa
Stage reaction	0.2

Determine the flow coefficient and stage loading factor at the mean radius and the reaction at the hub, radius 0.31 m, at the design speed of 8000 rev/min, given that the stage is to have a free vortex swirl at this speed. You may assume that losses are absent. Comment upon the results you obtain.

(Take $C_p = 1.148$ kJ/(kg °C) and $\gamma = 1.33$)

Solution. Sufficient data are given to solve the mean radius velocity triangles from which the flow coefficient and stage loading factors are obtained.

At $r = r_m = 0.36$ m, $\alpha_2 = 68$ deg, $p_2 = 207$ kPa, $R = 0.2$, $T_{o1} = T_{o2} = 838$ K and $p_{o1} = p_{o2} = 350$ kPa, assuming adiabatic frictionless nozzle flow. Since $h_{o2} = h_2 + \frac{1}{2}c_2^2$,

$$\begin{aligned} c_2^2 &= 2C_p(T_{o2} - T_2) \\ &= 2C_p T_{o2}\left[1 - (p_2/p_{o2})^{(\gamma-1)/\gamma}\right] \\ &= 2 \times 1148 \times 838\left[1 - (207/350)^{0.248}\right] = 23.51 \times 10^4 \end{aligned}$$

$$\therefore c_2 = 484.9 \text{ m/s}$$

The mean blade speed is

$$\begin{aligned} U_m &= (2\pi N/60)r_m = (2\pi \times 8000/60)0.36 \\ &= 301.6 \text{ m/s} \end{aligned}$$

Hence, the mean flow coefficient is

$$\begin{aligned} \phi_m &= c_x/U_m = c_2 \cos \alpha_2/U_m \\ &= 484.9 \times \cos 68^\circ/301.6 \end{aligned}$$

$$\therefore \underline{\phi_m = 0.6023}$$

The stage reaction is defined as

$$R = \frac{h_2 - h_3}{h_1 - h_3} = \frac{h_2 - h_3}{h_{o1} - h_{o3}} \qquad (\text{if } c_1 = c_3)$$

$$\therefore\ 1 - R = \frac{h_{o1} - h_{o3} - h_2 + h_3}{h_{o1} - h_{o3}} = \frac{c_2^2 - c_3^2}{2U(c_{\theta 2} + c_{\theta 3})} = \frac{c_{\theta 2} - c_{\theta 3}}{2U} \qquad \text{(i)}$$

At the mean radius,

$$c_{\theta 2} - c_{\theta 3} = 2U_m(1 - R_m) = 2 \times 301.6 \times 0.8 = 482.6\ \text{m/s},$$

$$c_{\theta 2} = c_2 \sin \alpha_2 = 484.9 \times \sin 68^{\circ} = 449.6\ \text{m/s}$$

$$\therefore\ c_{\theta 3} = -33.0\ \text{m/s}$$

The stage loading factor at the mean radius is

$$\psi_m = \Delta W/U_m^2 = (c_{\theta 2} + c_{\theta 3})/U_m = (449.6 - 33)/301.6$$

$$\therefore\ \underline{\psi_m = 1.381}$$

From eqn. (i) above, the reaction at any radius is

$$R = 1 - (c_{\theta 2} - c_{\theta 3})/(2U)$$

where, for a free-vortex, $c_{\theta 2} = K_2/r$, $c_{\theta 3} = K_3/r$ and the blade speed $U = \Omega r$.

Substituting for $c_{\theta 2}$, $c_{\theta 3}$ and U

$$R = 1 - k/(r/r_m)^2$$

where $k = (K_2 - K_3)/(2\Omega r_m^2)$

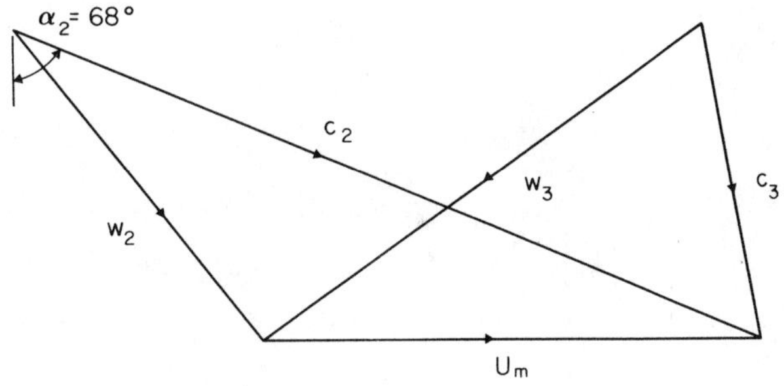

Velocity triangles at mean radius

Solving for k with R = 0.2 at $r = r_m$

$$R = 1 - 0.8/(r/r_m)^2$$

The reaction at the hub, $r = r_h = 0.31$ m, is

$$R_h = 1 - 0.8/0.861^2$$
$$= \underline{-0.079}$$

The negative reaction would imply that diffusion of the flow occurs in the rotor row (i.e. $w_3 < w_2$) at the root. For a turbine blade row, flow diffusion results in poor efficiency caused by large total pressure losses. A poor flow distribution will result and this can adversely affect the performance of any subsequent stages. Turbine designers always aim for a positive root reaction to avoid this problem.

6.3. Gas enters the nozzles of an axial flow turbine stage with uniform total pressure at a uniform velocity c_1 in the axial direction and leaves the nozzles at a constant flow angle α_2 to the axial direction. The absolute flow leaving the rotor c_3 is completely axial at all radii.

Using radial equilibrium theory and assuming no losses in total pressure show that

$$(c_3^2 - c_1^2)/2 = U_m c_{\theta m2} \left[1 - \left(\frac{r}{r_m} \right)^{\cos^2 \alpha_2} \right]$$

where U_m is the mean blade speed

$c_{\theta m2}$ is the tangential velocity component at nozzle exit at the mean radius $r = r_m$

(Note: The approximation $c_3 = c_1$ at $r = r_m$ is used to derive the above expression.)

Solution. This problem and the one following are both examples of the so-called 'direct problem' in which the flow angle is specified as some function of radius and the velocity component distributions $c_x(r)$ and $c_\theta(r)$ are to be solved.

In this question the essential idea to grasp is that as a result of the specific work variation with radius, the total pressure at stage outlet p_{o3} is also non-uniform. The equation for specific work done by a turbine stage, eqn. (4.2), applied to an incompressible, frictionless flow is

$$\Delta W = (p_{o2} - p_{o3})/\rho = (p_{o1} - p_{o3})/\rho = U c_{\theta 2} \qquad \text{(i)}$$

where $p_{o1} = p_1 + \frac{1}{2}\rho c_1^2$ and $p_{o3} = p_3 + \frac{1}{2}\rho c_3^2$. Noting that with no swirl in the flow at stage inlet and outlet, $dp/dr = 0$, and both p_1 and p_3 have constant values. Thus, from eqn. (i),

$$U c_{\theta 2} = \frac{1}{2}(c_1^2 - c_3^2) + k \qquad \text{(ii)}$$

where $k = (p_1 - p_3)/\rho$. From eqn. (6.22)

$$\frac{c_{\theta 2}}{c_{\theta m2}} = \left(\frac{r}{r_m}\right)^{-\sin^2 \alpha_2}$$

(a similar derivation of this equation is given in the solution of Q.6.4),

$$\therefore \quad \frac{U c_{\theta 2}}{U_m c_{\theta m2}} = \frac{r}{r_m} \frac{c_{\theta 2}}{c_{\theta m2}} = \left(\frac{r}{r_m}\right)^{\cos^2 \alpha_2} \qquad \text{(iii)}$$

Substituting eqn. (iii) into eqn. (ii),

$$U_m c_{\theta m2} \left(\frac{r}{r_m}\right)^{\cos^2 \alpha_2} = \frac{1}{2}(c_1^2 - c_3^2) + k$$

Noting that $c_1 = c_3$ at $r = r_m$, the constant $k = U_m c_{\theta m2}$. Hence,

$$(c_3^2 - c_1^2)/2 = U_m c_{\theta m2} \left[1 - \left(\frac{r}{r_m}\right)^{\cos^2 \alpha_2}\right]$$

6.4. Gas leaves an untwisted turbine nozzle at an angle α to the axial direction and in radial equilibrium. Show that the variation in axial velocity from root to tip, assuming total pressure is constant, is given by

$$c_x r^{\sin^2 \alpha} = \text{constant}.$$

Determine the axial velocity at a radius of 0.6 m when the axial velocity is 100 m/s at a radius of 0.3 m. The outlet angle α is 45 deg.

Solution. In an incompressible flow the total pressure is $p_o = p + \frac{1}{2}\rho c^2$. Differentiating this expression with respect to r and noting that p_o is assumed to be constant

$$\frac{1}{\rho}\frac{dp}{dr} + c\frac{dc}{dr} = 0 \qquad \text{(i)}$$

For a swirling flow in radial equilibrium

$$\frac{1}{\rho}\frac{dp}{dr} = \frac{c_\theta^2}{r} \qquad \text{(ii)}$$

Combining equations (i) and (ii),

$$\frac{c_\theta^2}{r} + c\frac{dc}{dr} = 0 \qquad \text{(iii)}$$

which is another form of the radial equilibrium equation. Noting that $c_\theta = c_x \tan \alpha$ and $c = c_x \sec \alpha$, substitution into eqn. (iii) gives,

$$\frac{c_x^2}{r}\tan^2\alpha + c_x\frac{dc_x}{dr}\sec^2\alpha = 0$$

After some simplification and re-arrangement to separate the variables,

$$\frac{dc_x}{c_x} = -\sin^2\alpha\frac{dr}{r}$$

which, for constant flow angle α, can be immediately integrated to give

$$\underline{c_x r^{\sin^2\alpha} = \text{constant}}$$

The numerical part is easily solved by direct substitution. At $r = 0.6$ m and with $\sin^2\alpha = 0.5$

$$c_x = 100(0.3/0.6)^{0.5}$$
$$= \underline{70.7 \text{ m/s}}$$

6.5. The flow at the entrance and exit of an axial flow compressor rotor is in radial equilibrium. The distributions of the tangential components of absolute velocity with radius are:

$$c_{\theta 1} = ar - b/r, \text{ before the rotor,}$$

$$c_{\theta 2} = ar + b/r, \text{ after the rotor,}$$

where a and b are constants. What is the variation of work done with radius? Deduce expressions for the axial velocity distributions before and after the rotor, assuming incompressible flow theory and that the radial gradient of stagnation pressure is zero.

At the mean radius, $r = 0.3$ m, the stage loading coefficient, $\psi = \Delta W/U_t^2$ is 0.3, the reaction ratio is 0.5 and the mean axial velocity is 150 m/s. The rotor speed is 7640 rev/min. Determine the rotor flow inlet and outlet angles at a radius of 0.24 m given that the hub/tip ratio is 0.5. Assume that at the mean radius the axial velocity remains unchanged ($c_{x1} = c_{x2}$ at r = 0.3 m).

(Note: ΔW is the specific work and U_t the blade tip speed.)

Solution. The specific work done on the gas, from eqn. (5.1), is

$$\Delta W = U(c_{\theta 2} - c_{\theta 1})$$

Substituting for the blade speed $U = \Omega r$ and the tangential velocity components

$$\Delta W = \Omega r(2b/r) = 2b\Omega = \text{constant.}$$

Thus, the work done is constant with radius. Note that for a uniform stagnation pressure at rotor inlet and ΔW = constant, the stagnation pressure after the rotor will be constant provided that the flow is either frictionless or that losses are distributed uniformly with radius.

From the radial equilibrium equation, eqn. (6.8), with p_0 = constant

$$\frac{d}{dr}(c_x^2/2) + \frac{c_\theta}{r}\frac{d}{dr}(r c_\theta) = 0$$

Considering the whirl distribution at rotor inlet, then

$$\frac{d}{dr}\left(\frac{c_{x1}^2}{2}\right) + (a - b/r^2)2ar = 0$$

Integrating this equation

$$\frac{c_{x1}^2}{2} = \int\left(\frac{2ab}{r} - 2a^2 r\right)dr = 2ab\,\ell n\,r - a^2 r^2 + k_1/2$$

Thus, the distributions in axial velocity at inlet and outlet of the rotor are, respectively,

$$c_{x1}^2 = k_1 - 2a^2\left[r^2 - (2b/a)\,\ell n\,r\right] \qquad \text{(i)}$$

$$c_{x2}^2 = k_2 - 2a^2\left[r^2 + (2b/a)\,\ell n\,r\right] \qquad \text{(ii)}$$

where k_1 and k_2 are arbitrary constants.

In order to determine the flow angles at any radius r it is necessary, first of all, to solve the four constants a, b, k_1 and k_2. The stage loading factor, which is constant for all radii because ΔW is constant, is defined as

$$\psi = \frac{\Delta W}{U_t^2} = \frac{(c_{\theta 2} - c_{\theta 1})_t}{U_t} = \frac{2b}{r_t U_t} = \frac{2b}{\Omega r_t^2} = 0.3$$

$$\therefore b = 0.15\,\Omega r_t^2 \qquad \text{(iii)}$$

The reaction ratio is defined, eqn. (5.11), as

$$R = c_x(\tan\beta_1 + \tan\beta_2)/(2U)$$

$$= 1 - (c_{\theta 1} + c_{\theta 2})/(2U)$$

At the mean radius r_m the reaction ratio is 0.5, hence

$$(c_{\theta 2} + c_{\theta 1})_m = U_m = 2 a r_m \qquad \text{(iv)}$$

after using the swirl equations given in the problem. Hence, with $U_m = \Omega r_m$,

$$a = \Omega/2 \qquad \text{(iva)}$$

The angular velocity, $\Omega = 2\pi N/60 = 2\pi \times 7640/60 = 800$ rad/s. The mean radius r_m is the arithmetic mean (other definitions of 'mean' are sometimes used!), i.e. $r_m = \frac{1}{2}(r_t + r_h)$, hence

$$r_t = 2r_m/(1 + r_h/r_t)$$

$$= 2 \times 0.3/(1 + 0.5)$$

$$= 0.4 \text{ m}$$

Thus, using eqns. (iva) and (iii), $a = 400$ and $b = 19.2$.

From eqns. (i) and (ii), with $c_{x1} = c_{x2} = 150$ m/s at $r = r_m = 0.3$ m,

$$k_1 = c_{x1}^2 + 2a^2\left[r^2 - 2(b/a)\,\ln r\right]$$

$$= 150^2 + 2 \times 400^2\left[0.3^2 - 0.096\,\ln 0.3\right] = 8.829 \times 10^4$$

$$k_2 = 150^2 + 2 \times 400^2\left[0.3^2 + 0.096\,\ln 0.3\right] = 1.431 \times 10^4$$

Hence, at $r = 0.24$ m,

$$c_{x1}^2 = 8.829 \times 10^4 - 32 \times 10^4\left[0.24^2 - 0.096\,\ln 0.24\right]$$

$$= 2.601 \times 10^4$$

$$\therefore c_{x1} = 161.3 \text{ m/s}$$

$$c_{x2}^2 = 1.431 \times 10^4 - 32 \times 10^4\left[0.24^2 + 0.096 \ln 0.24\right]$$

$$= 3.972 \times 10^4$$

$$\therefore c_{x2} = 199.3 \text{ m/s}$$

From the swirl distribution equations, at r = 0.24 m

$$c_{\theta 1} = ar - b/r = 400 \times 0.24 - 19.2/0.24 = 16 \text{ m/s}$$

$$c_{\theta 2} = ar + b/r = 400 \times 0.24 + 19.2/0.24 = 176 \text{ m/s}$$

For the axial compressor stage, the velocity triangles (Fig. 5.2) yield, at r = 0.24 m

$$\tan \beta_1 = (\Omega r - c_{\theta 1})/c_{x1} = (800 \times 0.24 - 16)/161.3 = 1.091$$

$$\tan \beta_2 = (\Omega r - c_{\theta 2})/c_{x2} = (800 \times 0.24 - 176)/199.3 = 0.0803$$

The relative flow angles at rotor inlet and outlet at r = 0.24 m are, respectively,

$$\underline{\beta_1 = 47.5 \text{ deg}}, \quad \underline{\beta_2 = 4.59 \text{ deg}}$$

6.6. An axial flow turbine stage is to be designed for free-vortex conditions at exit from the nozzle row and for zero swirl at exit from the rotor. The gas entering the stage has a stagnation temperature of 1000 K, the mass flow rate is 32 kg/s, the root and tip diameters are 0.56 m and 0.76 m respectively, and the rotor speed is 8000 rev/min. At the rotor tip the stage reaction is 50% and the axial velocity is constant at 183 m/s. The velocity of the gas entering the stage is equal to that leaving.

Determine:

(i) the maximum velocity leaving the nozzles;

(ii) the maximum absolute Mach number in the stage;

(iii) the root section reaction;

(iv) the power output of the stage;

(v) the stagnation and static temperatures at stage exit.

(Take R = 0.287 kJ/(kg $^{\circ}$C) and C_p = 1.147 kJ/(kg $^{\circ}$C)

Solution. A simplifying assumption which has been employed in deriving answers is that the axial velocity is constant both across the stage and radially.

(i) At nozzle exit, as $rc_{\theta 2} = K =$ constant, the velocity will be largest at the hub, radius $r = r_h = 0.28$ m. Now at the tip, radius $r = r_t = 0.38$ m, the reaction $R = 0.5$ (velocity triangles are symmetrical) and the leaving absolute velocity $c_3 = c_x = 183$ m/s. With $\Omega = 2\pi N/60 = \pi \times 8000/30 = 837.8$ rad/s, the blade tip speed $U_t = \Omega r_t = 318.4$ m/s. The tip section velocity diagram obtained from this data is shown below,

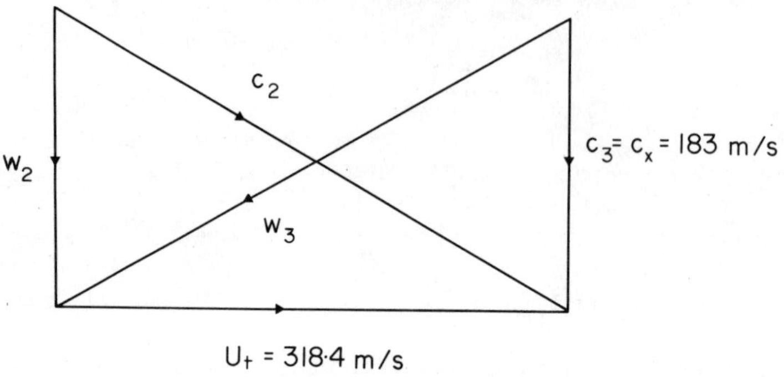

It is seen from the tip section velocity diagram that $c_{\theta 2t} = U_t$. Hence, $c_{\theta 2h} = c_{\theta 2t}\, r_t/r_h = 318.4 \times 0.38/0.28 = 432.1$ m/s. The maximum velocity at nozzle exit is,

$$c_{2h} = (c_{\theta 2h}^2 + c_x^2)^{1/2} = (432.1^2 + 183^2)^{1/2}$$

$$= \underline{469.3 \text{ m/s}}$$

(ii) The maximum absolute Mach number in the stage also occurs at nozzle exit at $r = r_h$ and is determined from

$$M_{2\,Max}^2 = c_{2h}^2/(\gamma R T_{2h})$$

where $T_{2h} = T_{o1} - c_{2h}^2/(2C_p) = 1000 - 469.3^2/(2 \times 1147) = 904.0$ K

and $\gamma = C_p/(C_p - R) = 1147/(1147 - 287) = 1.334$

$$\therefore M_{2\,Max} = 469.3/(1.334 \times 287 \times 904)^{1/2}$$

$$= \underline{0.798}$$

(iii) For a normal axial turbine stage (i.e. with $c_1 = c_3$) the degree of reaction, eqn. (4.22c), is

$$R = 1 + (c_{\theta 3} - c_{\theta 2})/(2U)$$

With $c_{\theta 3} = 0$ (i.e. axial exit flow), $U = \Omega r$ and $c_{\theta 2} = K/r$

$$\therefore R = 1 - K/(2\Omega r^2) = 1 - k/r^2$$

At the tip $r = r_t$, $R = R_t = 0.5$, hence $k = 0.5\, r_t^2$. Hence, the root section reaction is

$$R_h = 1 - 0.5\,(r_t/r_h)^2 = 1 - 0.5\,(0.38/0.28)^2$$

$$= \underline{0.079}$$

(iv) For a free-vortex turbine stage the specific work done is constant with radius. At the tip radius $c_{\theta 3} = 0$, $c_{\theta 2} = U_t$ and $\Delta W = U_t^2$. Thus, the power developed by the stage is

$$\dot{W}_t = \dot{m}\,\Delta W = \dot{m}\,C_p\,(T_{o1} - T_{o3}) = \dot{m}\,U_t^2 = 32 \times 318.4^2$$

$$= \underline{3.244 \text{ MW}}$$

(v) The stagnation and static temperatures at stage exit are

$$T_{o3} = T_{o1} - \Delta W/C_p = T_{o1} - U_t^2/C_p = 1000 - 318.4^2/1147$$

$$= 1000 - 88.39$$

$$= \underline{911.6 \text{ K}}$$

$$T_3 = T_{o3} - c_3^2/2C_p = 911.6 - 183^2/(2 \times 1147)$$

$$= \underline{897.0 \text{ K}}$$

6.7. The rotor blades of an axial flow turbine stage are 100 mm long and are designed to receive gas at an incidence of 3 deg from a nozzle row. A free-vortex whirl distribution is to be maintained between nozzle exit and rotor entry. At rotor exit the absolute velocity is 150 m/s in the axial direction at all radii. The deviation is 5 deg for the rotor blades and zero for the nozzle blades at all radii. At the hub, radius 200 mm, the conditions are as follows:

Nozzle outlet angle	70 deg
Rotor blade speed	180 m/s
Gas speed at nozzle exit	450 m/s

Assuming that the axial velocity of the gas is constant across the stage, determine

(i) the nozzle outlet angle at the tip;

(ii) the rotor blade inlet angles at hub and tip;

(iii) the rotor blade outlet angles at hub and tip;

(iv) the degree of reaction at root and tip.

Why is it essential to have a positive reaction in a turbine stage?

<u>Solution</u>. The gas flow angles are determined first from the velocity diagrams at the hub and the tip, then the blade angles are found by suitably correcting the flow angles for the incidence and deviation angles given. The velocity diagram for the hub is shown below :-

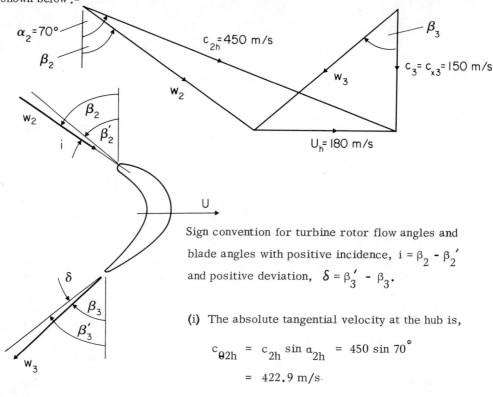

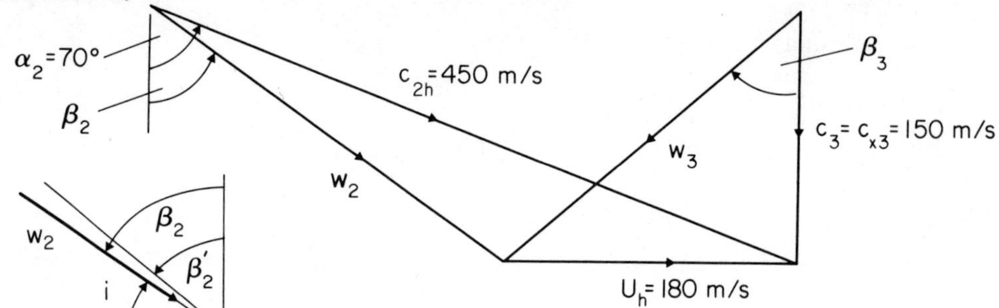

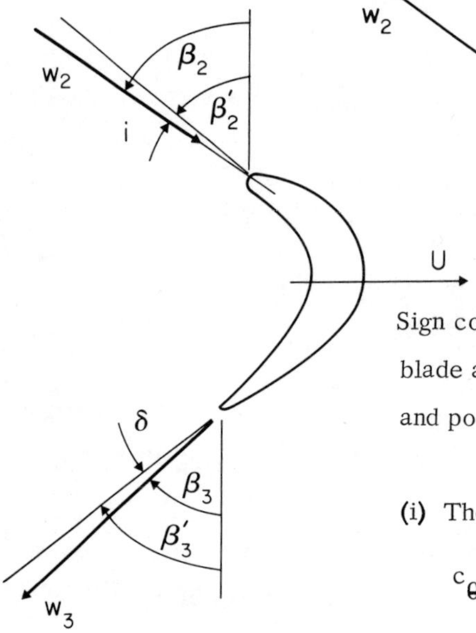

Sign convention for turbine rotor flow angles and blade angles with positive incidence, $i = \beta_2 - \beta_2'$ and positive deviation, $\delta = \beta_3' - \beta_3$.

(i) The absolute tangential velocity at the hub is,

$$c_{\theta 2h} = c_{2h} \sin \alpha_{2h} = 450 \sin 70^\circ$$
$$= 422.9 \text{ m/s}.$$

For the free-vortex flow, $r_t c_{\theta 2t} = r_h c_{\theta 2h}$

$$\therefore c_{\theta 2t} = c_{\theta 2h} r_h/r_t = 422.9 \times 2/3 = 281.9 \text{ m/s}$$

$$\therefore \tan \alpha_{2t} = c_{\theta 2t}/c_x = 281.9/150 = 1.879$$

$$\therefore \underline{\alpha_{2t} = \alpha_{2t}' = 62 \text{ deg},} \text{ as there is zero nozzle deviation.}$$

(ii) Referring to the velocity diagram, at the hub,

$$\tan\beta_{2h} = (c_{\theta 2h} - U_h)/c_x = (422.9 - 180)/150 = 1.619$$

$$\therefore \beta_{2h} = 58.3 \text{ deg.}$$

The rotor blade inlet angle at the hub is,

$$\beta_{2h}' = \beta_{2h} - i = 58.3 - 3$$

$$= \underline{55.3 \text{ deg}}.$$

Similarly, at the rotor tip,

$$\tan\beta_{2t} = (c_{\theta 2t} - U_t)/c_x = (281.9 - 180 \times 3/2)/150 = 0.0793$$

$$\therefore \beta_{2t} = 4.54 \text{ deg}$$

$$\therefore \beta_{2t}' = 4.54 - 3 = \underline{1.54 \text{ deg}}$$

(iii) Again, from the velocity diagram, at the hub,

$$\tan\beta_{3h} = U_h/c_x = 180/150 = 1.2$$

$$\therefore \beta_{3h} = 50.19 \text{ deg}$$

$$\therefore \beta_{3h}' = \beta_{3h} + \delta = \underline{55.19 \text{ deg}}$$

Similarly, at the tip

$$\tan\beta_{3t} = U_t/c_x = 1.5 \times 180/150 = 1.8$$

$$\beta_{3t} = 60.95 \text{ deg}$$

$$\therefore \underline{\beta_{3t}' = 65.95 \text{ deg}}$$

(iv) For a normal turbine stage (i.e. $c_1 = c_3$) with constant axial velocity across it, reaction is defined, eqn. (4.20), as

$$R = c_x(\tan\beta_3 - \tan\beta_2)/(2U)$$

At the hub,

$$R_h = 150(1.2 - 1.619)/360 = \underline{-0.175}$$

At the tip,

$$R_t = 150(1.8 - 0.079)/540 = \underline{0.478}$$

A positive degree of reaction in a turbine stage is necessary to avoid large total pressure losses caused by diffusion of the relative flow in the rotor.

N.B. There is a slight anomaly in the data given for this problem. At nozzle exit the axial velocity $c_{x2} = c_2 \cos \alpha_2 = 450 \cos 70 = 153.9$ m/s (at the hub) whereas the axial velocity is given as 150 m/s. This anomaly only slightly affects the numerical answers.

6.8. The rotor and stator of an isolated stage in an axial-flow turbomachine are to be represented by two actuator discs located at axial positions $x = 0$ and $x = \delta$ respectively. The hub and tip diameters are constant and the hub/tip radius ratio r_h/r_t is 0.5. The rotor disc considered on its own has an axial velocity of 100 m/s far upstream and 150 m/s downstream at a constant radius $r = 0.75\, r_t$. The stator disc in isolation has an axial velocity of 150 m/s far upstream and 100 m/s far downstream at radius $r = 0.75\, r_t$. Calculate and plot the axial velocity variation between $-0.5 \leqslant x/r_t \leqslant 0.6$ at the given radius for each actuator disc in isolation and for the combined discs when

$$\delta/r_t = 0.1, \quad 0.25 \quad \text{and} \quad 1.0.$$

Solution. The variation in axial velocity c_x with axial distance x away from an isolated actuator disc (located at $x = 0$ inside a cylindrical annulus) for a constant radial distance r is, eqn. (6.43), given to a first approximation by,

$$c_x = c_{x\infty 1} - \frac{1}{2}(c_{x\infty 1} - c_{x\infty 2}) \exp\left(\pi x/(r_t - r_h)\right), \quad x \leqslant 0 \qquad \text{(i)}$$

$$c_x = c_{x\infty 2} + \frac{1}{2}(c_{x\infty 1} - c_{x\infty 2}) \exp\left(-\pi x/(r_t - r_h)\right), \quad x \geqslant 0 \qquad \text{(ii)}$$

where $c_{x\infty 1}$, $c_{x\infty 2}$ are the axial velocities far upstream and far downstream respectively of the disc. For the rotor, $c_{x\infty 1} = 100$ m/s, $c_{x\infty 2} = 150$ m/s at $r/r_t = 0.75$ when $r_h/r_t = 0.5$. Substituting these values into the above equations

$$c_x = 100 + 25 \exp(2\pi x/r_t), \quad x \leqslant 0$$

$$c_x = 150 - 25 \exp(-2\pi x/r_t), \quad x \geqslant 0$$

Values of c_x are shown in tabular form below and graphically in Fig. Q.6.8(a):-

x/r_t	-0.5	-0.3	-0.2	-0.1	0	0.1	0.2	0.3	0.5
c_x m/s	101.1	103.8	107	113.4	125	136.7	143	146.2	148.9

The results for the isolated stator are not shown as they are merely the mirror image of those for the rotor with the origin shifted according to the stator's location.

When two actuator discs are in close proximity to one another, flow interference will occur, the magnitude of the effect being dependent upon the respective values of $c_{x\infty 1}$ and $c_{x\infty 2}$ for each disc. In the problem, the second disc representing the stator blade row is located at $x = \delta$. The far upstream axial velocity at $r/r_t = 0.75$ is the same as the far downstream axial velocity and is labelled $c_{x\infty 2}$. Similarly, the far downstream axial velocity of the stator disc at $r/r_t = 0.75$ is the same as the far upstream axial velocity of the rotor disc and is labelled $c_{x\infty 1}$. The axial velocity for the second disc in isolation is

$$c_x = c_{x\infty 2} - \frac{1}{2}(c_{x\infty 2} - c_{x\infty 1}) \exp[\pi(x - \delta)/(r_t - r_h)] \qquad \text{(iii)}$$

for $x \leqslant \delta$, and

$$c_x = c_{x\infty 1} + \frac{1}{2}(c_{x\infty 2} - c_{x\infty 1}) \exp[-\pi(x - \delta)/(r_t - r_h)] \qquad \text{(iv)}$$

for $x \geqslant \delta$. Substituting values for $c_{x\infty 1}$, $c_{x\infty 2}$ and r_h/r_t

$$c_x = 150 - 25 \exp[2\pi(x - \delta)/r_t], \quad x \leqslant \delta$$

$$c_x = 100 + 25 \exp[-2\pi(x - \delta)/r_t], \quad x \geqslant \delta$$

The axial velocity variation for the two discs in combination appropriate to each region, from eqns. (6.48), (6.49) and (6.50) are,

$$c_x = 100 + 25[\exp(2\pi x/r_t) - \exp(2\pi(x - \delta)/r_t)], \quad x \leqslant 0$$

$$c_x = 150 - 25[\exp(-2\pi x/r_t) + \exp(2\pi(x - \delta)/r_t)], \quad 0 \leqslant x \leqslant \delta$$

$$c_x = 100 + 25[\exp(-2\pi(x - \delta)/r_t) - \exp(-2\pi x/r_t)], \quad \delta \leqslant x$$

From the equations the variation in axial velocity for the three values of $\delta/r_t = 0.1$, 0.25 and 1.0 have been calculated and are shown graphically.

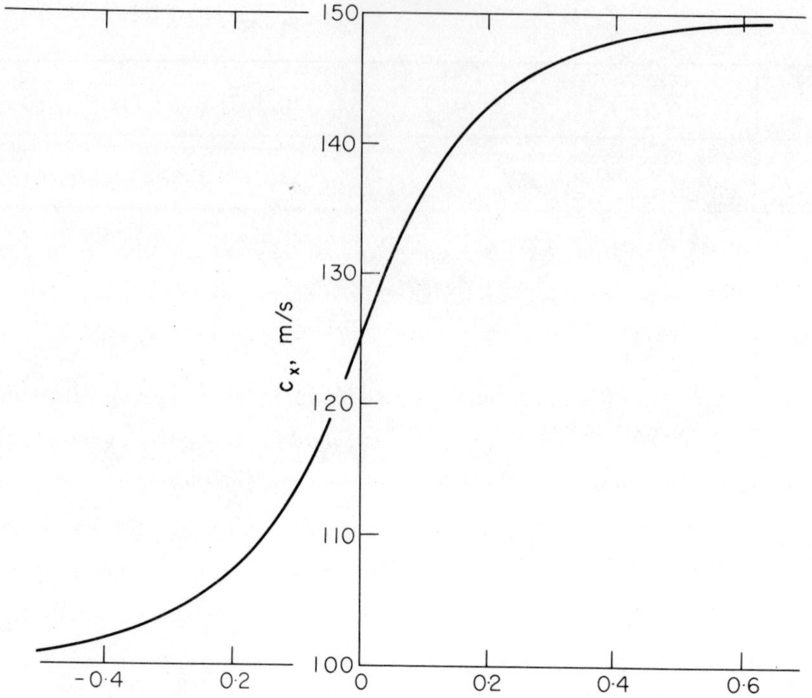

(a) Variation in axial velocity with axial distance from an isolated actuator disc located at x = 0, for $r/r_t = 0.75$

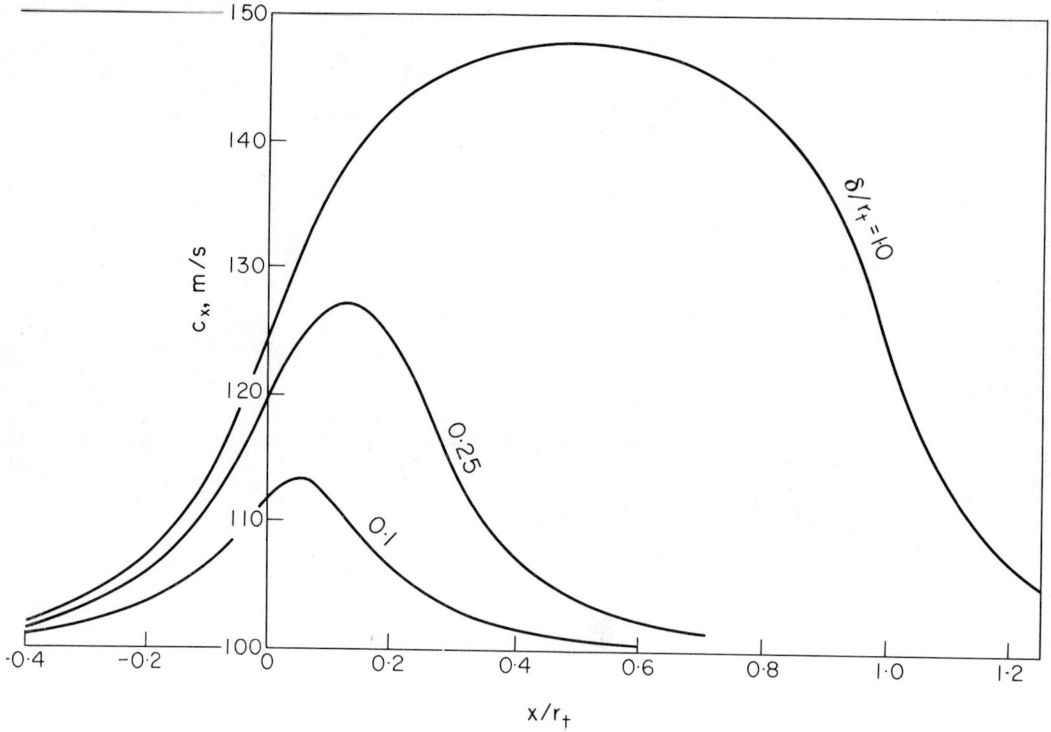

(b) Variation in axial velocity with axial distance from two actuator discs separated by a distance δ, one disc being located at the origin.

Chapter 7
Centrifugal Compressors and Pumps

Note. In problems 7.1 to 7.5 assume that the gas constant R = 287 J/(kg $^{\circ}$C) and that $\gamma = 1.4$. For problems 7.1 to 7.4 the stagnation pressure and stagnation temperature at compressor entry are assumed to be 101.3 kPa and 288 K respectively.

7.1. The air entering the impeller of a centrifugal compressor has an absolute axial velocity of 100 m/s. At rotor exit the relative air angle measured from the radial direction is 26° 36', the radial component of velocity is 120 m/s and the tip speed of the radial vanes is 500 m/s. Determine the power required to drive the compressor when the air flow rate is 2.5 kg/s and the mechanical efficiency is 95%. If the radius ratio of the impeller eye is 0.3, calculate a suitable inlet diameter assuming the inlet flow is incompressible. Determine the overall total pressure ratio of the compressor when the total-to-total efficiency is 80%, assuming the velocity at exit from the diffuser is negligible.

Solution. The specific work required to compress the air is $\Delta W = h_{03} - h_{01} = U_2 c_{\theta 2}$, the flow being without swirl at impeller entry. It follows from the velocity triangle at the impeller exit (see sketch) that

$$c_{\theta 2} = U_2 - c_{r2} \tan \beta_2$$

$$= 500 - 120 \times \tan 26.6^{\circ}$$

$$= 440 \text{ m/s}$$

$$\therefore \Delta W = 440 \times 500 = 220 \text{ kJ/kg}$$

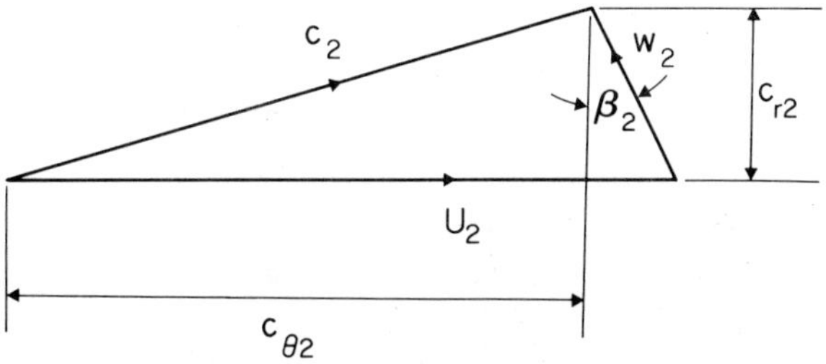

The theoretical power needed (i.e. ignoring mechanical losses) is

$$\dot{W}_c = \dot{m}\,\Delta W = 2.5 \times 220$$

$$= 550 \text{ kW}$$

Hence, the actual power needed is

$$P = \dot{W}_c/\eta_m = 550/0.95$$

$$= \underline{578.9 \text{ kW}}$$

The equation of continuity, with $c_{x1} = c_1$, is

$$\dot{m} = \rho_1 A_1 c_1 = \pi \rho_1 (r_{s1}^2 - r_{h1}^2) c_1$$
$$= \pi \rho_1 c_1 r_{s1}^2 \left[1 - (r_{h1}/r_{s1})^2\right]$$

With the assumption that the flow at entry is incompressible, the density ρ_1 can be determined from the stagnation pressure and temperature,

$$\rho_1 = \rho_{01} = p_{01}/(RT_{01}) = 1.013 \times 10^5/(287 \times 288)$$
$$= 1.226 \text{ kg/m}^3$$
$$\therefore r_{s1}^2 = \dot{m}/\{\pi \rho_1 c_1 \left[1 - (r_{h1}/r_{s1})^2\right]\}$$
$$= 2.5/(\pi \times 1.226 \times 100 \times 0.91)$$
$$= 0.7135 \times 10^{-2}$$
$$\therefore \underline{d_{s1} = 169 \text{ mm}}$$

The total to total efficiency of the compressor (eqn. 7.20) is

$$\eta_c = \frac{h_{03ss} - h_{01}}{h_{03} - h_{01}} = C_p T_{01} \left[(p_{03}/p_{01})^{(\gamma-1)/\gamma} - 1\right] / \Delta W$$

$$\therefore \frac{p_{03}}{p_{01}} = \left(1 + \frac{\eta_c \Delta W}{C_p T_{01}}\right)^{\gamma/(\gamma-1)} = \left(1 + \frac{0.8 \times 220 \times 10^3}{1005 \times 288}\right)^{3.5}$$
$$= 1.608^{3.5}$$
$$= \underline{5.273}$$

7.2. A centrifugal compressor has an impeller tip speed of 366 m/s. Determine the absolute Mach number of the flow leaving the radial vanes of the impeller when the radial component of velocity at impeller exit is 30.5 m/s and the slip factor is 0.90. Given that the flow area at impeller exit is 0.1 m^2 and the total-to-total efficiency of the impeller is 90%, determine the mass flow rate.

Solution. The absolute Mach number at impeller exit (eqn. 7.24) is

$$M_2 = c_2/a_2 = c_2/(\gamma RT_2)^{1/2}$$

so that c_2 and T_2 need to be determined. From the velocity triangle at impeller exit

$$c_2^2 = c_{\theta 2}^2 + c_{r2}^2 = (\sigma_s U_2)^2 + c_{r2}^2 = (0.9 \times 366)^2 + 30.5^2$$
$$= 1.094 \times 10^5$$

From the definitions of specific work (eqn. 7.1) and slip factor

$$\Delta W = \sigma_s U_2^2 = h_{o2} - h_{o1} = C_p(T_{o2} - T_{o1})$$
$$= C_p(T_2 - T_{o1}) + \frac{1}{2} c_2^2$$
$$\therefore T_{o2} = T_{o1} + \sigma_s U_2^2/C_p = 288 + 0.9 \times 366^2/1005 = 408 \text{ K}$$
$$\therefore T_2 = T_{o2} - \frac{1}{2} c_2^2/C_p = 408 - \frac{1}{2} \times 1.094 \times 10^5/1005 = 353.5 \text{ K}$$

Thus,

$$M_2 = \left[1.094 \times 10^5/(1.4 \times 287 \times 353.5)\right]^{1/2} = 0.8778$$

The rate of mass flow is $\dot{m} = \rho_2 A_2 c_{r2}$ where the density is $\rho_2 = p_2/(RT_2)$. The static pressure p_2 must be determined by first solving for the impeller total pressure ratio p_{o2}/p_{o1} and then relating p_2 to p_{o2} by means of the isentropic temperature-pressure equation. Thus, the impeller total to total efficiency is

$$\eta_i = \frac{T_{o2s} - T_{o1}}{T_{o2} - T_{o1}} = \frac{(p_{o2}/p_{o1})^{(\gamma-1)/\gamma} - 1}{T_{o2}/T_{o1} - 1}$$

$$\therefore \frac{p_{o2}}{p_{o1}} = \left(1 + \eta_i \left(\frac{T_{o2}}{T_{o1}} - 1\right)\right)^{3.5} = \left(1 + 0.9 \times \left(\frac{408}{288} - 1\right)\right)^{3.5} = 3.047$$

$$\frac{p_2}{p_{o2}} = \left(\frac{T_2}{T_{o2}}\right)^{\gamma/(\gamma-1)} = \left(\frac{353.5}{408}\right)^{3.5} = 1/1.154^{3.5} = 1/1.651$$

$$\therefore p_2 = \frac{p_2}{p_{o2}} \cdot \frac{p_{o2}}{p_{o1}} \cdot p_{o1} = \frac{3.047}{1.651} \times 10^5 = 184.6 \text{ kPa}$$

$$\dot{m} = p_2 A_2 c_{r2}/(RT_2) = 184.6 \times 10^3 \times 0.1 \times 30.5/(287 \times 353.5)$$
$$= \underline{5.550 \text{ kg/s}}$$

7.3. The eye of a centrifugal compressor has a hub/tip radius ratio of 0.4, a maximum relative flow Mach number of 0.9 and an absolute flow which is uniform and completely axial. Determine the optimum speed of rotation for the condition of maximum mass flow given that the mass flow rate is 4.536 kg/s. Also, determine the outside diameter of the eye and the ratio of axial velocity/blade speed at the eye tip. Figure 7.4 may be used to assist the calculations.

Solution. The compressible flow relation between mass flow rate, speed of rotation and the flow parameters at the eye tip is given by eqn. (7.11),

$$0.6598 \times 10^{-8} \frac{\dot{m}\Omega^2}{k} = \frac{M_{r1}^2 \sin^2 \beta_{s1} \cos \beta_{s1}}{(1 + 0.2 M_{r1}^2 \cos^2 \beta_{s1})^4}$$

when $\gamma = 1.4$, $R = 287$ J/(kg ^{o}C), $p_{01} = 101.3$ kPa and $T_{01} = 288$ K. Now rather than using the appropriate curve in Fig. 7.4 to determine the maximum mass flow condition, greater accuracy is obtained by differentiating the RHS of the above equation with respect of $\cos \beta_{s1}$ and setting the result to zero. Putting $x = \cos \beta_{s1}$, $M_{r1} = 0.9$ and differentiating etc., the following equation is found,

$$(1 + 0.162\, x^2)(1 - 3x^2) = 1.296x^2(1 - x^2)$$

$$\therefore 0.81x^4 - 4.134x^2 + 1 = 0$$

Solving this quadratic equation, the only valid root is

$$x^2 = 0.2546$$

$$\therefore \cos \beta_{s1} = \sqrt{0.2546} = 0.5046$$

$$\therefore \beta_{s1} = 59.7 \text{ deg}$$

Hence,

$$\Omega^2 = \frac{0.729 \times 10^8 \times \sin^2 59.7^{\circ} \times \cos 59.7^{\circ} \times 0.84}{0.6598 \times 4.536(1 + 0.162 \times \cos^2 59.7)^4}$$

$$= 6.547 \times 10^6 (\text{rad/s})^2$$

$$\therefore \Omega = 2559 \text{ rad/s} = \underline{24,430 \text{ rev/min}}$$

From the equation of continuity, the rate of mass flow is

$$\dot{m} = \rho_1 A_1 c_{x1} = \pi \rho_1 k r_{s1}^2 c_{x1}$$

where $k = 1 - 0.4^2 = 0.84$ and ρ_1 (at the relatively high Mach number prevailing) is sought from compressible flow theory. From the inlet velocity triangle (see sketch),

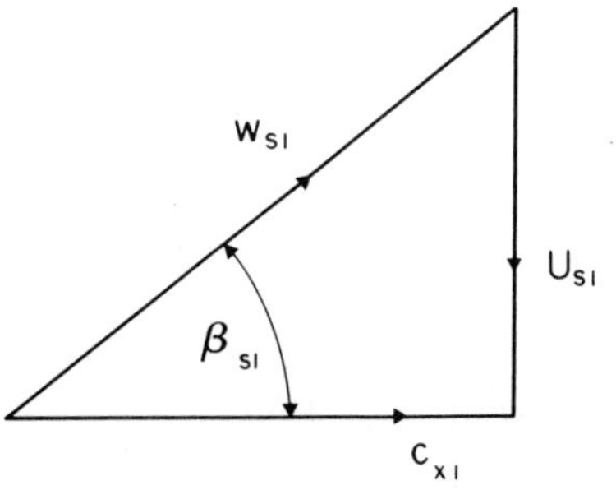

$$w_{s1} = c_{x1} \sec \beta_{s1}$$

Also $M_{r1} = w_{s1}/a_1$ and $M_1 = c_{x1}/a_1$

$$\therefore M_1 = M_{r1} \cos \beta_{s1} = 0.9 \times 0.5046$$
$$= 0.4541$$

Thus, using the compressible flow relation,

$$T_{01}/T_1 = 1 + \frac{1}{2}(\gamma - 1)M_1^2 = 1 + 0.2 \times 0.4541^2 = 1.0412$$

$$c_{x1} = M_1 a_1 = M_1\sqrt{(\gamma R T_1)}$$
$$= 0.4541(1.4 \times 287 \times 288/1.0412)^{1/2}$$
$$= 151.4 \text{ m/s}$$

$$\rho_1 = \rho_{01}(T_1/T_{01})^{1/(\gamma-1)} = \rho_{01}(1/1.0412)^{2.5} = \rho_{01}/1.106$$

Thus,

$$r_{s1}^2 = \frac{\dot{m}}{\pi \rho_1 k c_{x1}} = \frac{\dot{m} R T_{01} \times 1.106}{\pi p_{01} k c_{x1}}$$

$$= \frac{4.536 \times 287 \times 288 \times 1.106}{\pi \times 1.013 \times 10^5 \times 0.84 \times 151.4} = 1.024 \times 10^{-2}$$

$$\therefore \underline{d_{s1} = 202.4 \text{ mm}}$$

The ratio of axial velocity to blade tip speed at the eye is

$$c_{x1}/U_{s1} = \cot \beta_{s1} = \underline{0.5844}$$

7.4. An experimental centrifugal compressor is fitted with free-vortex guide vanes in order to reduce the relative air speed at inlet to the impeller. At the outer radius of the eye, air leaving the guide-vanes has a velocity of 91.5 m/s at 20 deg to the axial direction. Determine the inlet relative Mach number, assuming frictionless flow through the guide vanes, and the impeller total-to-total efficiency.

Other details of the compressor and its operating conditions are:

Impeller entry tip diameter, 0.457 m
Impeller exit tip diamter, 0.762 m
Slip factor 0.9
Radial component of velocity at impeller exit, 53.4 m/s
Rotational speed of impeller, 11,000 rev/min
Static pressure at impeller exit, 223 kPa (abs.)

Solution. The inlet guide vanes reduce the inlet relative velocity w_1 by deflecting the absolute flow through an angle α_1, producing a tangential velocity $c_{\theta 1}$ in the same direction as the blade motion (see sketch of velocity triangle at inlet). The required relative Mach number, which will be greatest at the shroud, is defined as,

$$M_{1rel} = w_1/a_1$$

where, from the inlet velocity triangle

$$w_1^2 = (U_{s1} - c_{\theta 1})^2 + c_{x1}^2,$$

$$a_1^2 = \gamma R T_1$$

$$T_1 = T_{o1} - \frac{1}{2} c_1^2/C_p$$

$$= 288 - \frac{1}{2} \times 91.5^2/1005$$

$$= 283.8 \text{ K}$$

$$\therefore a_1 = (1.4 \times 287 \times 283.8)^{1/2}$$

$$= 337.7 \text{ m/s}$$

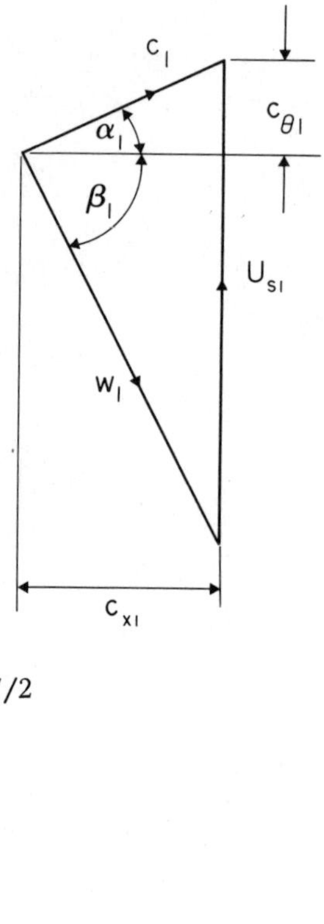

The inlet blade tip speed is

$$U_{s1} = \Omega r_{s1} = 11000 \times (\pi/30) \times 0.457/2$$

$$= 263.2 \text{ m/s}$$

$$c_{\theta 1} = c_1 \sin \alpha_1 = 91.5 \times \sin 20^{\circ}$$

$$= 31.3 \text{ m/s}$$

$$\therefore w_1 = \left[U_{s1}^2 - 2 U_{s1} c_{\theta 1} + c_1^2\right]^{1/2}$$

$$= \left[263.2^2 - 2 \times 263.2 \times 31.3 + 91.5^2\right]^{1/2}$$

$$= 247.3 \text{ m/s}$$

$$\therefore M_{1rel} = 247.3/337.7 = \underline{0.7324}$$

The impeller total to total efficiency is (see Sol. Q.7.2)

$$\eta_i = \frac{(p_{o2}/p_{o1})^{(\gamma-1)/\gamma} - 1}{T_{o2}/T_{o1} - 1}$$

From eqn. (7.1) the specific work is

$$\Delta W = C_p(T_{o2} - T_{o1}) = U_2 c_{\theta 2} - U_1 c_{\theta 1}$$

$$\therefore \frac{T_{o2}}{T_{o1}} - 1 = \frac{\sigma U_2^2 - U_1 c_{\theta 1}}{C_p T_{o1}}, \quad U_2 = (r_2/r_{s1})U_{s1} = (0.762/0.457)\,263.2$$

$$= 438.9 \text{ m/s}$$

$$= (0.9 \times 438.9^2 - 263.2 \times 31.3)/(1005 \times 288)$$

$$= 0.5704$$

The impeller total pressure ratio is

$$\frac{p_{o2}}{p_{o1}} = \frac{p_2}{p_{o1}} \cdot \frac{p_{o2}}{p_2} = \frac{p_2}{p_{o1}} \left(\frac{T_{o2}}{T_2}\right)^{\gamma/(\gamma-1)}$$

Now $T_{o2} = T_2 + \frac{1}{2} c_2^2/C_p$

$$\therefore T_2/T_{o2} = 1 - c_2^2/(2C_p T_{o2}) = 1 - (c_{\theta 2}^2 + c_{r2}^2)/(2C_p T_{o2})$$

$$c_{\theta 2} = \sigma U_2 = 0.9 \times 438.9 = 395 \text{ m/s}$$

$$T_{o2} = 1.5704 \times 288 = 452.3 \text{ K}$$

$$\therefore T_2/T_{o2} = 1 - (395^2 + 53.4^2)/(2 \times 1005 \times 452.3) = 0.8252$$

$$\eta_i = \frac{(T_{o2}/T_2)(p_2/p_{o1})^{(\gamma-1)/\gamma} - 1}{T_{o2}/T_{o1} - 1}$$

$$= \frac{(1/0.8252)(223/101.3)^{1/3.5} - 1}{0.5704} = \underline{0.9084}$$

7.5. A centrifugal compressor has an impeller with 21 vanes, which are radial at exit, a vaneless diffuser and no inlet guide vanes. At inlet the stagnation pressure is 100 kPa abs. and the stagnation temperature is 300 K.

(i) Given that the mass flow rate is 2.3 kg/s, the impeller tip speed is 500 m/s and the mechanical efficiency is 96%, determine the driving power on the shaft. Use eqn. (7.18a) for the slip factor.

(ii) Determine the total and static pressures at diffuser exit when the velocity at that position is 100 m/s. The total to total efficiency is 82%.

(iii) The reaction, which may be defined as for an axial flow compressor by eqn. (5.10b), is 0.5, the absolute flow speed at impeller entry is 150 m/s and the diffuser efficiency is 84%. Determine the total and static pressures, absolute Mach number and radial component of velocity at the impeller exit.

(iv) Determine the total-to-total efficiency for the impeller.

(v) Estimate the inlet/outlet radius ratio for the diffuser assuming the conservation of angular momentum.

(vi) Find a suitable rotational speed for the impeller given an impeller tip width of 6 mm.

Solution. (i) For a radial vaned impeller, $\beta_2' = 0$ and the slip factor (eqn. 7.18a) is

$$\sigma_s = 1 - 0.63\pi/21 = 0.9057$$

The shaft power, $P = \dot{W}_c/\eta_m = \dot{m}(h_{o3} - h_{o1})/\eta_m$

$$h_{o3} - h_{o1} = U_2 c_{\theta 2} = \sigma_s U_2^2 = 0.9057 \times 500^2 = 226.4 \text{ kJ/kg}$$

$$\therefore P = 2.3 \times 226.4/0.96 = \underline{542.5 \text{ kW}}$$

(ii) The total to total efficiency of the centrifugal compressor is

$$\eta_c = \frac{h_{o3ss} - h_{o1}}{h_{o3} - h_{o1}} = \frac{C_p T_{o1}(T_{o322}/T_{o1} - 1)}{\sigma_s U_2^2}$$

$$\frac{T_{o3ss}}{T_{o1}} = \left(\frac{p_{o3}}{p_{o1}}\right)^{(\gamma-1)/\gamma}, \quad C_p T_{o1} = \frac{a_{o1}^2}{\gamma-1}$$

$$\therefore \frac{p_{o3}}{p_{o1}} = \left[1 + (\gamma - 1)\eta_c \sigma_s U_2^2/a_{o1}^2\right]^{\gamma/(\gamma-1)}$$

$$= \left[1 + \frac{0.4 \times 0.82 \times 2.264 \times 10^5}{1.4 \times 287 \times 300}\right]^{3.5}$$

$$= 5.365$$

$$\therefore p_{o3} = \underline{536.5 \text{ kPa}}$$

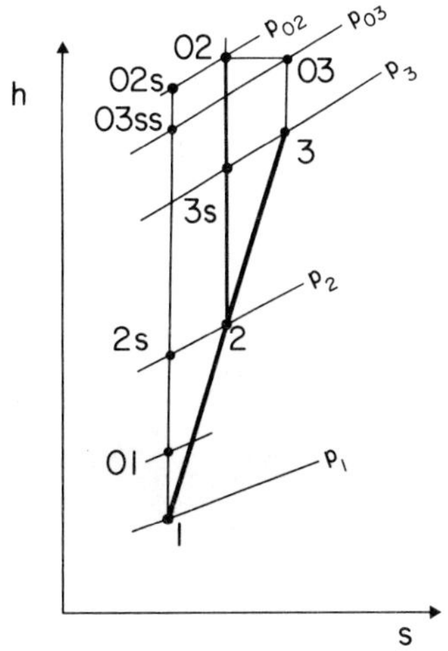

$$\frac{p_{o3}}{p_3} = \left(\frac{T_{o3}}{T_3}\right)^{\gamma/(\gamma-1)}$$

$$= \left[\frac{T_{o3}}{T_{o3} - c_3^2/(2C_p)}\right]^{\gamma/(\gamma-1)}$$

$$= 1/\left[1 - c_3^2/(2h_{o3})\right]^{\gamma/(\gamma-1)}$$

$$h_{o3} = (h_{o3} - h_{o1}) + h_{o1}$$

$$= (226.4 + 1.005 \text{ x } 300)10^3$$

$$= 527.9 \text{ kJ/kg}$$

$$\therefore p_3 = 536.5\left[1 - 100^2/(2 \text{ x } 52.79 \text{ x } 10^4)\right]^{3.5} = \underline{518.9 \text{ kPa}}$$

(iii) At impeller exit the absolute Mach number $M_2 = c_2/a_2$ where the speed of sound $a_2 = \sqrt{\{(\gamma - 1)h_2\}}$. The enthalpy h_2 and velocity c_2 are evaluated as follows. From the definition of reaction $R = (h_2 - h_1)/(h_3 - h_1) = 0.5$ and

$$h_3 - h_1 = h_{o3} - h_{o1} + \frac{1}{2}(c_1^2 - c_3^2) = \left[22.64 + \frac{1}{2}(2.25 - 1)\right] \text{ x } 10^4$$

$$= 232.7 \text{ kJ/kg}$$

$$\therefore h_2 - h_1 = \frac{1}{2}(h_3 - h_1) = 116.4 \text{ kJ/kg}$$

$$\text{Now } h_2 = (h_{o1} - \frac{1}{2}c_1^2) + (h_2 - h_1) = (301.5 - 11.25 + 116.4) \text{ x } 10^3$$

$$= 406.7 \text{ kJ/kg}$$

$$c_2^2 = 2(h_{o3} - h_2) \quad \text{since } h_{o2} = h_{o3}$$

$$= 2\left[(h_{o3} - h_{o1}) + (h_{o1} - h_2)\right]$$

$$= 2\left[226.4 + 301.5 - 406.7\right] \text{ x } 10^3 = 242.4 \text{ x } 10^3$$

Thus,

$$M_2 = \left[242.4/(0.4 \text{ x } 406.7)\right]^{1/2} = \underline{1.221}$$

The diffuser efficiency is defined (eqn. 5.10b) as

$$\eta_d = \frac{h_{3s} - h_2}{h_3 - h_2} = \frac{h_2(T_{3s}/T_2 - 1)}{h_3 - h_2} = \frac{h_2\left[(p_3/p_2)^{(\gamma-1)/\gamma} - 1\right]}{h_3 - h_2}$$

$$\therefore \frac{p_3}{p_2} = \left[1 + \eta_d(h_3 - h_2)/h_2\right]^{\gamma/(\gamma-1)}$$

$$= \left[1 + 0.84 \times 116.4/406.7\right]^{3.5} = 2.126$$

$$\therefore p_2 = 518.9/2.126 = \underline{244.1 \text{ kPa}}$$

$$\frac{p_{o2}}{p_2} = \left(\frac{T_{o2}}{T_2}\right)^{\gamma/(\gamma-1)} = \left(\frac{h_{o3}}{h_2}\right)^{\gamma/(\gamma-1)} = \left[\frac{527.9}{406.7}\right]^{3.5} = 2.492$$

$$\therefore p_{o2} = \underline{608.2 \text{ kPa}}$$

From the impeller tip velocity triangle,

$$c_{r2}^2 = c_2^2 - c_{\theta 2}^2 = c_2^2 - (\sigma_s U_2)^2$$

$$= 242.4 \times 10^3 - (0.9057 \times 500)^2 = 37.33 \times 10^3$$

$$c_{r2} = \underline{193.2 \text{ m/s}}$$

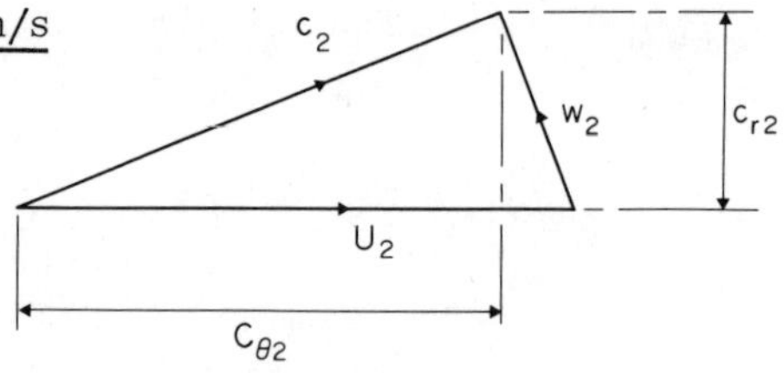

(iv) For the impeller, the total to total efficiency is

$$\eta_i = \frac{h_{o2s} - h_{o1}}{h_{o2} - h_{o1}} = \frac{h_{o1}\left[(p_{o2}/p_{o1})^{(\gamma-1)/\gamma} - 1\right]}{h_{o3} - h_{o1}}$$

$$= 301.5(6.082^{1/3.5} - 1)/226.4$$

$$= \underline{0.899}$$

(v) Assuming angular momentum is conserved, rc_θ = constant, and

$$\frac{r_2}{r_3} = \frac{c_{\theta 3}}{c_{\theta 2}} \doteqdot \frac{c_3}{\sigma_s U_2} = \frac{100}{0.9057 \times 500} = \underline{0.2208}$$

(vi) The angular velocity of the impeller is $\Omega = U_2/(2\pi r_2)$. The impeller tip radius is found from the equation of continuity, $\dot{m} = \rho_2 A_2 c_{r2}$, where $A_2 = 2\pi r_2 b_2$, and the equation of state. Thus,

$$\rho_2 = \frac{p_2}{RT_2} = \frac{\gamma p_2}{(\gamma-1)h_2} = \frac{3.5 \times 244.1 \times 10^3}{406.7 \times 10^3} = 2.101 \text{ kg/m}^3$$

$$\Omega = U_2 \rho_2 b_2 c_{r2}/\dot{m} = 500 \times 2.101 \times 0.006 \times 193.2/2.3$$

$$= 529.5 \text{ rev/s}$$

$$= \underline{31,770 \text{ rev/min}}$$

7.6. A centrifugal pump is used to raise water against a static head of 18.0 m. The suction and delivery pipes, both 0.15 m diameter, have respectively, friction head losses amounting to 2.25 and 7.5 times the dynamic head. The impeller, which rotates at 1450 rev/min, is 0.25 m diameter with 8 vanes, radius ratio 0.45, inclined backwards at $\beta_2' = 60$ deg. The axial width of the impeller is designed so as to give constant radial velocity at all radii and is 20 mm at impeller exit. Assuming an hydraulic efficiency of 0.82 and an overall efficiency of 0.72, determine

(i) the volume flow rate;
(ii) the slip factor using Busemann's method;
(iii) the impeller vane inlet angle required for zero incidence angle;
(iv) the power required to drive the pump.

Solution. (i) The actual head H which is delivered by the pump is defined as the difference in head measured between the outlet and inlet flanges of the pump. It is equal to the static head H_s, defined as the difference in level between the two open reservoirs, plus all external head losses. The external losses comprise the friction losses in the suction and delivery pipes together with the kinetic energy leaving the delivery pipe. Thus,

$$H = H_s + (2.25 + 7.5 + 1)c^2/(2g)$$

where the average velocity in the pipes, both of diameter d, is

$$c = 4Q/(\pi d^2) = 4Q/(\pi \times 0.15^2) = 56.59Q \text{ m/s}$$

$$\therefore H = 18 + 10.75 \times (56.59Q)^2/19.62 = 18 + 1755Q^2 \text{ m}$$

Now the ideal head $H_i = U_2 c_{\theta 2}/g$ and the hydraulic efficiency of the pump is defined as

$$\eta_h = \frac{H}{H_i} = \frac{gH}{U_2 c_{\theta 2}} = \frac{gH}{U_2^2 \sigma_B (c_{\theta 2}/U_2)}$$

The Busemann slip factor σ_B favoured in most pump design calculations, is

$$\sigma_B = c_{\theta 2}/c'_{\theta 2} = (A - B\phi_2 \tan\beta'_2)/(1 - \phi_2 \tan\beta'_2)$$

where A and B are constants determined by the geometry of a particular pump, β'_2 is the impeller vane outlet angle and $\phi_2 = c_{r2}/U_2$. The vane tip speed U_2 is

$$U_2 = \pi N D_2/60 = \pi \times 1450 \times 0.25/60 = 18.98 \text{ m/s}$$

$$\therefore \phi_2 = Q/(\pi D_2 b_2 U_2) = 4Q/(\pi \times 0.02 \times 18.98) = 3.351\,Q$$

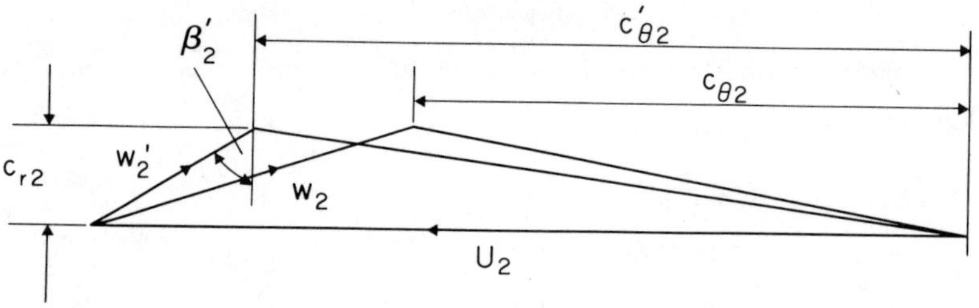

For this pump the radius ratio $r_2/r_1 = 2.222$ is sufficiently large (i.e.> exponential $(2\pi \cos\beta'_2/Z) = 1.481$, see eqn. 7.17c) for B to be assumed equal to unity and A can be found by interpolating from the graph shown, which was obtained from Fig. 7.10. Thus, at $\beta'_2 = 60$ deg and $Z = 8$ the value of $B = 0.818$ with sufficient accuracy. After substitution

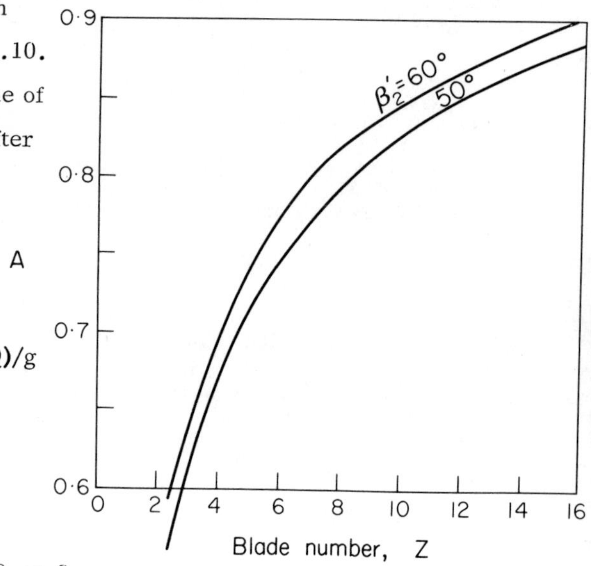

$$\eta_h = \frac{gH}{U_2^2 (0.818 - 1.732\,\phi_2)}$$

$$H = \eta_h U_2^2 (0.818 - 3.351 \times 1.732Q)/g$$

$$18 + 1755Q^2 = 0.82 \times 18.98^2 \times (0.818 - 5.804Q)/9.81$$

After simplifying,

$$Q^2 + 0.0996Q - 0.00378 = 0$$

$$\therefore Q = 0.02932 \text{ m}^3/\text{s}$$
$$= \underline{29.32 \text{ dm}^3/\text{s}}$$

(ii) With the volumetric flow rate solved the Busemann slip factor is easily obtained.

Now $\phi_2 = 3.351 \times 0.0293 = 0.09825$

$$\therefore \sigma_B = (0.818 - 0.09825 \times 1.732)/(1 - 0.09825 \times 1.732)$$
$$= 0.6478/0.8298$$
$$= \underline{0.7807}$$

(iii) The impeller vane inlet angle β_1' for zero flow incidence is obtained from

$$\cot \beta_1' = c_{r1}/U_1 = (c_{r2}/U_2)(r_2/r_1) = \phi_2 r_2/r_1$$
$$= 0.09825/0.45 = 0.2183$$

as $c_{r1} = c_{r2} = c_r$ for this pump. Thus,

$$\beta_1' = \underline{77.68 \text{ deg}}$$

(iv) The power required is

$$\dot{W}_p = \rho Q g H/\eta_h$$
$$= \rho Q g (18 + 1755 Q^2)/\eta_h$$
$$= 29.32 \times 9.81 \times 19.51/0.72 = \underline{7.794 \text{ kW}}$$

Chapter 8

Radial Flow Turbines

8.1. A small inward flow radial gas turbine, comprising a ring of nozzle blades, a radial vaned impeller and an axial diffuser, operates at its design point with a total to total efficiency of 0.90. At turbine entry the stagnation pressure and temperature of the gas is 400 kPa and 1,140 K. The flow leaving the turbine is diffused to a pressure of 100 kPa and has negligible final velocity. Given that the flow is just choked at nozzle exit, determine the impeller peripheral speed and the flow outlet angle from the nozzles.

For the gas assume $\gamma = 1.333$ and $R = 287$ J/(kg $^{\circ}$C).

Solution. The figure below shows a meridional section of a 90 deg inward flow radial turbine and diffuser together with the design point velocity triangles at rotor inlet and rotor outlet. At this condition the relative velocity w_2 at rotor inlet is in the radial direction and at rotor outlet the absolute velocity c_3 is in the axial direction (i.e. $c_{\theta 2} = U_2$ and $c_{\theta 3} = 0$). Thus, the specific work, eqn. (8.4), is

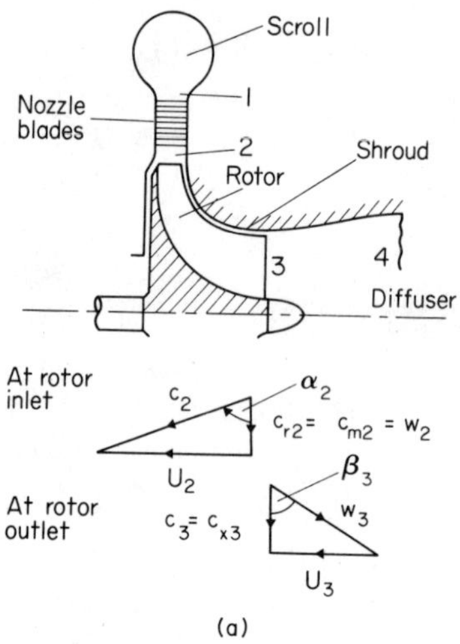

(a)

$$\Delta W = h_{o1} - h_{o3}$$

$$= U_2 c_{\theta 2} - U_3 c_{\theta 3} = U_2^2$$

Referring to the simplified Mollier diagram shown below, the total to total efficiency of the combined turbine stage and diffuser is

$$\eta_{tt} = (h_{o1} - h_{o4})/(h_{o1} - h_{o4ss}) = (h_{o1} - h_{o3})/(h_{o1} - h_{o4ss})$$

$$= U_2^2 / \left[C_p T_{o1} (1 - T_{o4ss}/T_{o1})\right]$$

After transposing and substituting for the isentropic temperature T_{o4ss}/T_{o1} in terms of the pressure ratio,

$$U_2^2 = \eta_{tt} C_p T_{o1}\left[1 - (p_{o4}/p_{o1})^{(\gamma-1)/\gamma}\right]$$
$$= 0.9 \times 1149 \times 1140\left[1 - (100/400)^{0.2498}\right]$$
$$= 1.179 \times 10^6 (1 - 0.7073) = 0.3451 \times 10^6$$

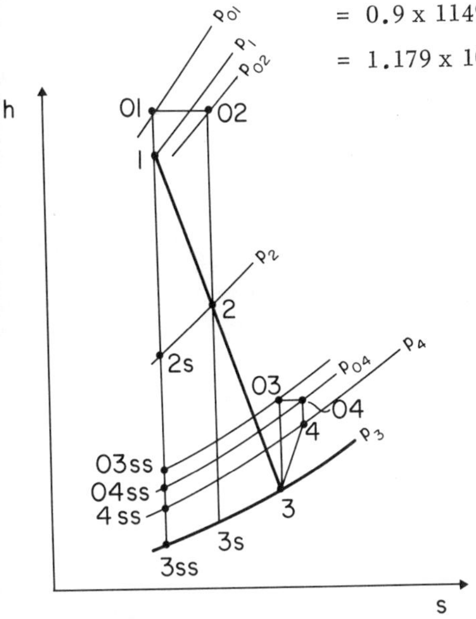

Hence, the blade tip speed is

$$U_2 = \underline{587.4 \text{ m/s}}.$$

At nozzle exit the absolute flow Mach number is

$$M_2 = c_2/a_2$$
$$= (U_2/a_2) \operatorname{cosec} \alpha_2 \qquad \text{(i)}$$

In eqn. (i) the values of both a_2 and α_2 are unknown and another condition must be used to solve the flow angle α_2. Across the nozzle the stagnation enthalpy remains constant, i.e.

$$h_{o1} = h_{o2}$$
$$\therefore C_p T_{o1} = C_p T_2 + \frac{1}{2} c_2^2 = C_p T_2 + \frac{1}{2} U_2^2 \operatorname{cosec}^2 \alpha_2$$

After rearranging,

$$T_2/T_{o1} = 1 - \frac{1}{2} U_2^2 \operatorname{cosec}^2 \alpha_2/(C_p T_{o1})$$
$$= 1 - \frac{1}{2}(\gamma - 1)(U_2/a_{o1})^2 \operatorname{cosec}^2 \alpha_2 \qquad \text{(ii)}$$

where $\quad C_p T_{o1} = \gamma R T_{o1}/(\gamma - 1) = a_{o1}^2/(\gamma - 1)$

Using eqn. (i)

$$T_2/T_{o1} = (a_2/a_{o1})^2 = U_2^2 \operatorname{cosec}^2 \alpha_2/(M_2 a_{o1})^2 \qquad \text{(iii)}$$

Combining eqns. (ii) and (iii) and rearranging

$$\sin \alpha_2 = (U_2/a_{o1})\left(\frac{1}{2}(\gamma - 1) + 1/M_2^2\right)^{1/2} \qquad \text{(iv)}$$

Substituting values,

$$a_{o1} = (\gamma R T_{o1})^{1/2} = (1.333 \times 287 \times 1140)^{1/2}$$

$$= 660.4 \text{ m/s}$$

$$\therefore \sin \alpha_2 = (587.4/660.4)\left(1 + 0.1665\right)^{1/2} = 0.9607$$

Hence, the nozzle flow outlet angle is

$$\alpha_2 = \underline{73.88^{\circ}}$$

8.2. The mass flow rate of gas through the turbine given in Problem 8.1 is 3.1 kg/s, the ratio of the impeller axial width/impeller tip radius (b_2/r_2) is 0.1 and the nozzle isentropic velocity ratio (ϕ_2) is 0.96. Assuming that the space between nozzle exit and impeller entry is negligible and ignoring the effects of blade blockage, determine:

(i) the static pressure and static temperature at nozzle exit;

(ii) the impeller tip diameter and rotational speed;

(iii) the power transmitted assuming a mechanical efficiency of 93.5%.

Solution. (i) The static temperature T_2 at nozzle exit can be obtained immediately from either eqn. (ii) or eqn. (iii) in the solution of Problem 8.1. Using eqn. (iii),

$$T_2 = T_{01}\, U_2^2/(M_2\, a_{01} \sin \alpha_2)^2$$

$$= 1140 \times 587.4^2/(660.4 \times 0.9607)^2$$

$$= \underline{977.2 \text{ K}}$$

The static pressure p_2 at nozzle exit may be found using the nozzle isentropic velocity ratio, eqn. (8.17),

$$\phi_2 = c_2/c_{2s}$$

to determine the isentropic temperature ratio T_{2s}/T_{01} as follows:-

$$h_{01} - h_{2s} = \frac{1}{2}c_{2s}^2, \quad h_{01} - h_2 = \frac{1}{2}c_2^2$$

$$\therefore \phi_2^2 = (h_{01} - h_2)/(h_{01} - h_{2s}) = (T_{01} - T_2)/(T_{01} - T_{2s})$$

After some rearranging,

$$T_{2s}/T_{01} = 1 - (1 - T_2/T_{01})/\phi_2^2 = 1 - (1 - 977.2/1140)/0.96^2$$

$$= 0.8450$$

$$p_2/p_{01} = (T_{2s}/T_{01})^{\gamma/(\gamma-1)} = 0.845^{4.003} = 0.5096$$

$$\therefore p_2 = 0.5096 \times 400$$

$$= \underline{203.8 \text{ kPa}}$$

(ii) Applying the equation of continuity at nozzle exit

$$\dot{m} = \rho_2 A_2 c_{r2}$$

where, from the design point velocity triangle,

$$c_{r2} = U_2 \cot \alpha_2$$

and the flow area is

$$A_2 = 2\pi r_2 b_2.$$

Hence, using the gas law,

$$r_2^2 = \frac{\dot{m} R T_2 \tan \alpha_2}{2\pi p_2 U_2 (b_2/r_2)} = \frac{3.1 \times 287 \times 977.2 \times \tan 73.88^o}{2\pi \times 203.8 \times 10^3 \times 587.4 \times 0.1}$$

$$= 0.03999$$

$$\therefore r_2 = 0.20 \text{ m}$$

$\therefore$ The impeller tip diameter is 40 cm.

The speed of rotation is

$$N = 60\, U_2/(\pi D_2) = 60 \times 587.4/(\pi \times 0.4)$$

$$= \underline{28,050 \text{ rev/min}}$$

(iii) The power transmitted taking into account the mechanical losses is

$$\dot{W}_t = \eta_m \dot{m} \Delta W = \eta_m \dot{m} U_2^2 = 0.935 \times 3.1 \times 587.4^2$$

$$= \underline{1,000 \text{ kW}}$$

8.3. A radial turbine is proposed as the gas expansion element of a nuclear powered Brayton cycle space power system. The pressure and temperature conditions through the stage at the design point are to be as follows:

Upstream of nozzles, $p_{o1} = 699$ kPa, $T_{o1} = 1,145$ K;

Nozzle exit, $p_2 = 527.2$ kPa, $T_2 = 1,029$ K;

Rotor exit, $p_3 = 384.7$ kPa, $T_3 = 914.5$ K, $T_{o3} = 924.7$ K.

The ratio of rotor exit mean diameter to rotor inlet tip diameter is chosen as 0.49 and the required rotational speed as 24,000 rev/min. Assuming the relative flow at rotor inlet is radial and the absolute flow at rotor exit is axial, determine:

(i) the total to static efficiency of the turbine;

(ii) the rotor diameter;

(iii) the implied enthalpy loss coefficients for the nozzles and rotor row.

The gas employed in this cycle is a mixture of helium and xenon with a molecular weight of 39.94 and a ratio of specific heats of 5/3. The Universal gas constant is, $R_o = 8.314$ kJ/(kg-mol K).

<u>Solution.</u> (i) The total to static efficiency of a radial flow turbine is defined, eqn. (8.6), as

$$\begin{aligned}
\eta_{ts} &= (h_{o1} - h_{o3})/(h_{o1} - h_{3ss}) \\
&= (1 - T_{o3}/T_{o1})/(1 - T_{3ss}/T_{o1}) \\
&= (1 - T_{o3}/T_{o1})/\left[1 - (p_3/p_{o1})^{(\gamma-1)/\gamma}\right] \\
&= (1 - 924.7/1145)/\left[1 - (384.7/699)^{0.4}\right] \\
&= (1 - 0.8076)/(1 - 0.7875) \\
&= \underline{0.9054}
\end{aligned}$$

(ii) At the design condition the relative flow at rotor inlet is radial and, from the velocity triangle (Fig. 8.3), $c_{\theta 2} = U_2$. As the absolute flow at rotor exit is axial, $c_{\theta 3} = 0$, and the specific work $\Delta W = U_2^2$.

$$\therefore U_2^2 = h_{o1} - h_{o3} = C_p(T_{o1} - T_{o3}) \qquad \text{(i)}$$

The Universal gas constant $R_o = Rm$ where m is the molecular 'weight' of the gas mixture. Thus,

$$\begin{aligned}
R &= R_o/m = 8314/39.94 = 208.2\ \text{J/(kg}\,^{\circ}\text{C)} \\
C_p &= \gamma R/(\gamma - 1) = 2.5 \times 208.2 = 520.5\ \text{J/(kg}\,^{\circ}\text{C)} \\
\therefore U_2^2 &= 520.5\,(1145 - 924.7) = 11.47 \times 10^4
\end{aligned}$$

$$U_2 = 338.6 \text{ m/s}$$

Hence, the rotor tip diameter is

$$D_2 = 60U_2/(\pi N) = 60 \times 338.6/(\pi \times 2.4 \times 10^4)$$
$$= \underline{0.2694 \text{ m}}$$

(iii) The enthalpy loss coefficient for the nozzles, eqn. (8.16), is determined as follows:-

$$\zeta_N = (h_2 - h_{2s})/(\tfrac{1}{2}c_2^2) = (T_2 - T_{2s})/(T_{01} - T_2)$$
$$= \left[T_2 - (T_{2s}/T_{01})T_{01}\right]/(T_{01} - T_2)$$
$$= \left[T_2 - T_{01}(p_2/p_{01})^{(\gamma-1)/\gamma}\right]/(T_{01} - T_2)$$
$$= \left[1029 - 1145\,(527.2/699)^{0.4}\right]/(1145 - 1029)$$
$$= (1029 - 1022.8)/(1145 - 1029)$$
$$= \underline{0.05316}$$

The enthalpy loss coefficient for the rotor is defined, eqn. (8.20), as

$$\zeta_R = (h_3 - h_{3s})/(\tfrac{1}{2}w_3^2) \qquad \text{(ii)}$$

The enthalpy loss $h_3 - h_{3s}$ is determined using the specified static pressures and temperatures,

$$h_3 - h_{3s} = C_p\left[T_3 - (T_{3s}/T_2)T_2\right] = C_p\left[T_3 - T_2(p_3/p_2)^{(\gamma-1)/\gamma}\right]$$
$$= 520.5\left[914.5 - 1029\,(384.7/527.2)^{0.4}\right]$$
$$= 3.831 \text{ kJ/kg}$$

From the rotor exit velocity triangle

$$w_3^2 = c_3^2 + U_3^2$$

where $U_3 = U_2(r_3/r_2)$ and $c_3^2 = 2\,C_p(T_{03} - T_3)$. Hence,

$$w_3^2 = 2C_p(T_{03} - T_3) + U_2^2(r_3/r_2)^2$$

Substituting for U_2^2 using eqn. (i), the relative kinetic energy at rotor exit is,

$$\tfrac{1}{2}w_3^2 = \left[C_p \;\; T_{03} - T_3 + \tfrac{1}{2}(r_3/r_2)^2(T_{01} - T_{03})\right]$$

$$= 520.5\left[924.7 - 914.5 + \frac{1}{2} \times 0.49^2(1145 - 924.7)\right]$$

$$= 19.07 \text{ kJ/kg}$$

Substituting these values into eqn. (ii),

$$\zeta_R = 3.831/19.07 = \underline{0.2009}$$

8.4. A film-cooled radial inflow turbine is to be used in a high performance open Brayton cycle gas turbine. The rotor is made of a material able to withstand a temperature of 1145 K at a tip speed of 600 m/s for short periods of operation. Cooling air is supplied by the compressor which operates at a stagnation pressure ratio of 4 to 1, with an adiabatic efficiency of 80%, when air is admitted to the compressor at a stagnation temperature of 288 K. Assuming that the effectiveness of the film cooling is 0.30 and the cooling air temperature at turbine entry is the same as that at compressor exit, determine the maximum permissible gas temperature at entry to the turbine.

Take $\gamma = 1.4$ for the air. Take $\gamma = 1.333$ for the gas entering the turbine. Assume $R = 287$ J/(kg K) in both cases.

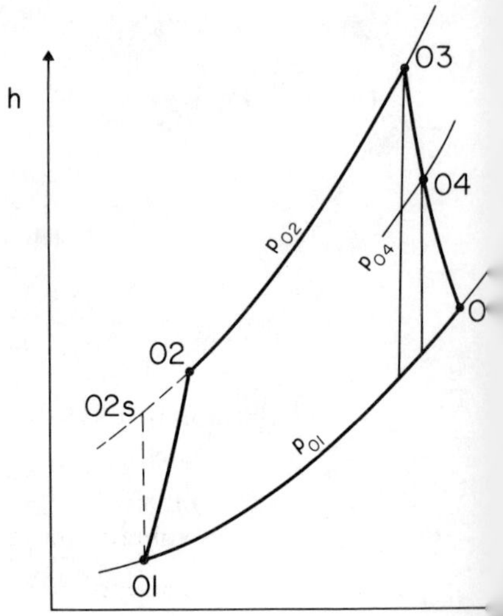

Solution. The sketch shows a Brayton gas turbine cycle in the form of a Mollier diagram. The compressor raises the stagnation pressure of the air from p_{01} to p_{02} at an adiabatic (total to total) efficiency η_c of 0.80. From the definition of efficiency, eqn. (2.28a),

$$\eta_c = (h_{02s} - h_{01})/(h_{02} - h_{01})$$

$$= T_{01}(T_{02s}/T_{01} - 1)/(T_{02} - T_{01})$$

$$= T_{01}\left[(p_{02}/p_{01})^{(\gamma-1)/\gamma} - 1\right]/(T_{02} - T_{01})$$

Hence, the actual stagnation temperature at compressor exit is,

$$T_{02} = T_{01}\left[1 + \left\{(p_{02}/p_{01})^{(\gamma-1)/\gamma} - 1\right\}/\eta_c\right]$$

$$= 288\left[1 + (4^{1/3.5} - 1)/0.8\right] = 288\left[1 + 0.486/0.8\right]$$

$= 463.0$ K

The cooling effectiveness ε, eqn. (8.41), in the present notation is

$$\varepsilon = \left[T_{o3} - T_m - \Delta W/(2C_p)\right] / \left[T_{o3} - T_{o2} - \Delta W/(2C_p)\right]$$

where T_m is the maximum permissible temperature of the rotor material and ΔW is the specific work done, assumed to be equal to U_2^2. Hence, after some rearrangement

$$T_{o3} = (T_m - \varepsilon T_{o2})/(1 - \varepsilon) + U_2^2/(2C_p)$$

where $C_p = \gamma R/(\gamma - 1) = 4.003 \times 287 = 1149$ J/(kg $^{\circ}$C)

$$\therefore T_{o3} = (1145 - 0.3 \times 463)/0.7 + 600^2/(2 \times 1149)$$

$$= \underline{1594 \text{ K}}$$

The specific work done by the turbine, $\Delta W_t = U_2^2 = 360$ kJ/kg of gas admitted to the radial turbine. The specific work required by the compressor, $\Delta W_c = C_p(T_{o2} - T_{o1})$ $= 175.9$ kJ/kg of air compressed (i.e. only about half of ΔW_t). This is feasible if it is considered that the rate of mass flow through the compressor is about twice that which flows through the turbine, the excess compressed air being used for other purposes. Only about 10% of this diverted compressed air would be needed for film cooling. At outlet from the turbine the hot gases would still have a stagnation pressure p_{o4} in excess of p_{o1}. Assuming the turbine has an adiabatic efficiency of 0.9 it is easy to show that $p_{o2}/p_{o4} = 2.682$ and, therefore, p_{o4}/p_{o1} is 1.492. This excess pressure ratio would be expanded through a second turbine stage to provide a net power output.

8.5. The radial inflow turbine in Problem 8.3 is designed for a specific speed Ω_s of 0.55 (rad). Determine,

(i) the volume flow rate at rotor exit and hence obtain the turbine power output,

(ii) the rotor exit hub and tip diameters,

(iii) the nozzle exit flow angle and the rotor inlet passage width/diameter ratio, b_2/D_2.

Solution. (i) Specific speed is defined, c.f. eqn. (8.33), as

$$\Omega_s = \Omega Q_3^{1/2}/\Delta h_{os}^{3/4}$$

where Q_3 is the volume flow rate at rotor exit, (m^3/s), Ω is the angular speed of the rotor (rad/s) and Δh_{os} is the isentropic enthalpy drop between inlet stagnation conditions and exhaust static conditions, $h_{o1} - h_{3ss}$. This last definition conforms with the use of total to static efficiency in Problem 8.3. Thus,

$$\Delta h_{os} = \frac{1}{2} c_o^2 = h_{o1} - h_{3ss} = C_p(T_{o1} - T_{3ss})$$

$$= C_p T_{o1}\left[1 - (p_3/p_{o1})^{(\gamma-1)/\gamma}\right]$$

$$= 520.5 \times 1145\left[1 - (384.7/699)^{0.4}\right] = 126.6 \text{ kJ/kg}$$

$$\Omega = 2\pi N/60 = 800\pi = 2513 \text{ rad/s}$$

Hence, from the specific speed definition,

$$Q_3^{1/2} = (\Omega_s/\Omega)\,\Delta h_{os}^{3/4} = (0.55/2513)(12.66 \times 10^4)^{3/4} = 1.469$$

$$\therefore Q_3 = \underline{2.159 \text{ m}^3/\text{s}}$$

The turbine power is

$$\dot{W}_t = \dot{m}\,\Delta W = \dot{m} U_2^2 = \rho_3 Q_3 U_2^2$$

$$\rho_3 = p_3/(RT_3) = 384.7 \times 10^3/(208.2 \times 914.5) = 2.021 \text{ kg/m}^3$$

$$\therefore \dot{m} = \rho_3 Q_3 = 2.021 \times 2.159 = 4.363 \text{ kg/s}$$

$$\therefore \dot{W}_t = 4.363 \times 338.6^2 = \underline{500 \text{ kW}}$$

(ii) $$Q_3 = A_3 c_3 = (\pi/4)(D_{3t}^2 - D_{3h}^2)c_3 = (\pi/2)D_{3m} H c_3$$

where,

$$H = D_{3t} - D_{3h},\quad D_m = (D_{3t} + D_{3h})/2 = 0.49 \times D_2 = 0.132 \text{ m}$$

$$c_3^2 = 2C_p(T_{o3} - T_3) = 2 \times 520.5\,(924.7 - 914.5) = 10620$$

$$\therefore c_3 = 103 \text{ m/s}$$

Hence, the rotor exit blade height is

$$H = 2Q_3/(\pi D_m c_3) = 2 \times 2.159/(\pi \times 0.132 \times 103)$$

$$= 0.1011 \text{ m}$$

$$\therefore D_{3h} = D_{3m} - H/2 = 0.132 - 0.0506 = \underline{0.0814 \text{ m}}$$

$$\therefore D_{3t} = D_{3m} + H/2 = \underline{0.1826 \text{ m}}$$

The rotor exit hub/tip ratio $D_{3h}/D_{3t} = 0.4458$ and the ratio $D_{3t}/D_2 = 0.678$.

(iii) As $h_{o1} = h_{o2}$, then

$$c_2^2 = 2C_p(T_{o1} - T_2) = 2 \times 520.5(1145 - 1029) = 12.08 \times 10^4$$

The nozzle exit velocity is

$$c_2 = 347.5 \text{ m/s}$$

From the inlet velocity triangle

$$\sin \alpha_2 = U_2/c_2 = 338.6/347.5 = 0.97439$$

$$\therefore \underline{\alpha_2 = 77.0^o}$$

From the equation of continuity,

$$\dot{m} = \rho_2 A_2 c_{r2} = \pi \rho_2 (b_2/D_2) D_2^2 U_2 \cot \alpha_2$$

$$\therefore b_2/D_2 = (\dot{m} R T_2 \tan \alpha_2)/(\pi p_2 D_2^2 U_2)$$

$$= \frac{4.363 \times 208.2 \times 1029 \times \tan 77^o}{\pi \times 527.2 \times 10^3 \times 0.2694^2 \times 338.6}$$

$$= \underline{0.0995}$$